Interference-Corrected Explicitly-Correlated Second-Order Perturbation Theory

Zur Erlangung des akademischen Grades eines
Doktors der Naturwissenschaften

(Dr. rer. nat.)

der Fakultät für Chemie und Biowissenschaften des
Karlsruher Instituts für Technologie
angenommene
Dissertation

von

M.Sc.

Konstantinos D. Vogiatzis

aus Piräus,
Griechenland

Dekan:	Prof. Dr. M. Bastmeyer
Referent:	Prof. Dr. W. Klopper
Korreferent:	PD Dr. Karin FInk

Tag der mündlichen Prüfung: 17. Oktober 2012

Konstantinos D. Vogiatzis

INTERFERENCE-CORRECTED EXPLICITLY-CORRELATED SECOND-ORDER PERTURBATION THEORY

ibidem-Verlag
Stuttgart

Bibliografische Information der Deutschen Nationalbibliothek
Die Deutsche Nationalbibliothek verzeichnet diese Publikation in der
Deutschen Nationalbibliografie; detaillierte bibliografische Daten sind im
Internet über http://dnb.d-nb.de abrufbar.

Bibliographic information published by the Deutsche Nationalbibliothek
Die Deutsche Nationalbibliothek lists this publication in the Deutsche Nationalbibliografie;
detailed bibliographic data are available in the Internet at http://dnb.d-nb.de.

Cover picture: © copyright 2012 by Marilena D. Vogiatzi

∞

Gedruckt auf alterungsbeständigem, säurefreien Papier
Printed on acid-free paper

ISBN-13: 978-3-8382-0477-2

© *ibidem*-Verlag
Stuttgart 2013

Printed in Germany

Στους γονείς μου, Ιλεάνα και Διονύσιο

Contents

CHAPTER 1

Introduction

It was in 1998 when John Pople was awarded the Nobel Prize in Chemistry "for his development of computational methods in quantum chemistry". In his famous lecture [1], he demonstrated in an absolute manner how the exact solution of the non-relativistic electronic Schrödinger equation can be obtained. There are two main parameters for this target: the size of the basis set chosen, i.e. the set of functions which will form the many-body wave function, and the level of excitations which is included in a computational method. The **exact** solution is calculated in an infinite (complete) set of functions where all combinations of electronic excitations (single, double, triple, etc. excitations) are considered. But why are these two parameters so important?

The basis set are used for the representation of the molecular orbitals, i.e. the probability of finding an electron in any specific region. These are one-electron expansions for a N-electron model. The independent molecular orbitals and the non occupied by electrons (virtual) orbitals are formulated from the solution of the Hartree-Fock equations. These unoccupied orbitals constitute the virtual space where the electron excitations are taking place. Mathematically, they consist of the higher eigenfunctions of the Fock operator. A complete basis set (CBS) provides the exact wave function and therefore a limitless space for excitations. Practically, the use of such an infinite number of functions is not possible. What is done in most computational chemistry programs is to start with a truncated, finite basis. The more terms that are included, the higher the accuracy that is achieved. And this is exactly what is shown in the x-axis of Figure 1.1.

The second crucial parameter is the level of excitations of electrons from their ground state into the virtual space. The more functions in the basis set, the larger the virtual space will be. They are important because they are used for the calculation of the correlation energy of the molecule under study. The correlation energy arises from the approximate representation of electron-electron repulsion in the Hartree-Fock (HF) method [1]. These repulsions are obtained from the instantaneous interactions (correlations) between electrons which can be moved (excited) into the virtual space. As it can be seen from Figure 1.1, HF is located in the bottom of the y axis. Above HF are some of the most common correlated methods used nowadays. Second-order Møller-Plesset perturbation theory (MP2) adds electron correlation to the HF energy from double excitations. Coupled-cluster theory [2] with iterative single-and-double excitations (CCSD) includes extra terms than MP2. The CCSD method is more accurate but more computationally demanding. The CCSD(T) method [3], that is above the CCSD level in Figure 1.1, includes a pertubative treatment of the triples excitations to the fifth-order The vital contribution of the "(T)" for higher accuracy is nowadays fully clear [4,5] and this method considered the *"golden standard"* of quantum chemistry. If all possible excitations are included, then the full configuration-interaction (FCI) level of theory is reached. FCI with a CBS means the **exact** solution of the non-relativistic electronic Schrödinger equation is calculated.

---

[1]IUPAC Gold Book

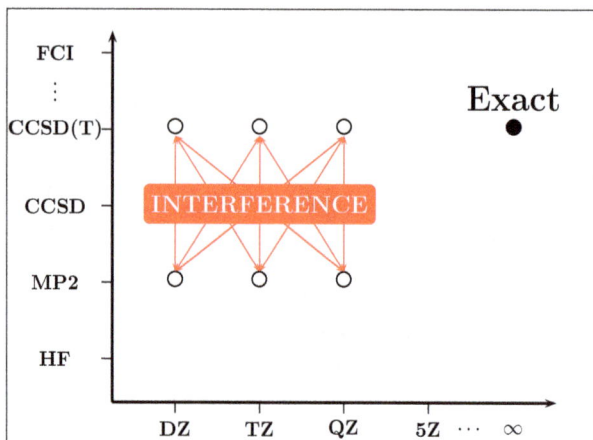

Figure 1.1: Schematic representation of the interference effects. The black dot stands for the basis set limit of the CCSD(T) method, target of the method presented in this Thesis.

Unfortunately, FCI/CBS calculations, *desirable in principle*, are usually too costly to apply except for very small systems. The CCSD(T) energy at the basis set limit (shown with a black dot in Figure 1.1) is adequate to describe most chemical applications and procedures. CCSD(T) with large basis sets, i.e. quintuple-zeta quality (5Z) or larger, are also not feasible for "real chemistry" applications. It also remains very time demanding to execute calculations with the smaller basis sets (TZ or QZ), despite the increase of computational power and the development of efficient codes. The reason for this drawback is the fact that the iterative singles and doubles need N^6 computational steps, while the pertubative triples are calculated once, but with a computational step of N^7, where N corresponds to the number of the basis functions. A variety of different approaches that speed-up the CCSD(T) method without loss of accuracy have been proposed, such as explicitly-correlated (F12) methods [6] and extrapolation schemes [7].

Explicitly-correlated methods expand, with extra (pair) functions, the virtual space in such a manner that they include *explicitly* the behavior of two electrons that approach very close to one another (electron cusp). Normally, this behavior is adequately described by a very large basis set but F12 methods recover this with smaller basis sets. The price to be paid in terms of computational time for this advantage is relatively small but the gain in accuracy, for the same basis set, is significantly high. [8] Nowadays, F12 methods have become a useful tool, not only for benchmark calculations [9, 10], but also for "real chemistry" applications [11, 12]. On the other hand, extrapolation schemes are based on the mathematical, asymptotic form of the correlation energy. These schemes can estimate with high accuracy the CBS by using smaller, truncated basis sets. Extrapolate and conquer! From the above we conclude that both approaches reach a higher level of accuracy whilst using smaller sets of functions. This accuracy is typically obtained from conventional methods but with (extremely) large basis sets and considerable computational effort.

But what is the accuracy that can be reached when both approaches, explicitly-correlated methods and extrapolation schemes, are combined? The idea of merging these techniques is the central focus of this Thesis. The advantages of the explicitly-correlated methods are incorporated in the extrapolation scheme proposed by Petersson and co-workers [13]. The new method formulated and implemented provides an estimate to the CCSD(T) method at the CBS limit. It uses as a starting point the CCSD(T) energy calculated with a truncated basis set and shifts it to the basis set limit by using interference effects (INT) from explicitly-correlated second-order Møller-Plesset perturbation theory (MP2-F12). Therefore, it is termed as CCSD(T)-INT-F12. The interference term arises from Petersson's suggestion that the first-order wave function *interferes* with the second-order (MP2) basis set truncation error. This error is calculated from the difference between the conventional and the explicitly-correlated MP2 theory. The red arrows in Figure 1.1 show schematically these effects: MP2 and CCSD(T) energies obtained from finite basis sets provide the basis set limit of the CCSD(T) theory (black dot). Vertical arrows represent the case where both CCSD(T) and MP2 energies are obtained with the same basis, while diagonal arrows stand for combinations of different bases (dual-basis). In addition, the method provides significant savings in computational time and requirements (disk space and memory).

The main goal of the present work is to present the capabilities of the CCSD(T)-INT-F12 method and the accuracy which can be achieved from it. In Chapter 2 the theoretical background of the interference effects and the MP2-F12 theory will be given. Technical details on the implementation and the performance of the CCSD(T)-INT-F12 method are presented in Chapter 3 with examples that verify the theory. The performance of the method has been tested in two different applications which are analyzed in separate Chapters. The first is associated with thermochemistry (Chapter 4) and the second is related with noncovalent interacting systems (Chapter 5). These applications are typical examples of common calculations executed by computational chemists. The highly accurate results obtained from the CCSD(T)-INT-F12 level of theory demonstrate in a complete manner the capabilities of the newly proposed method.

CHAPTER 2

Towards the Complete Basis Set Limit

2.1 Extrapolation Schemes

In molecular electronic structure theory, an as nearly complete as possible expansion of the molecular electronic wave function in one- and n-electron space is required for a highly accurate calculation. The expansion of the electronic wave function in orbital products (i.e. Slater determinants) leads to a frustratingly slow convergence to the basis set limit. This slow convergence remains the main obstacle for highly accurate calculations, as far as the computational cost increases much faster and unequally with the desirable accuracy while, at the same time, truncation of the atomic orbital (AO) basis leads to significant errors.

There are two different approaches for solving this problem. The first approach is concerned with the asymptotic form of the series. Based on the convergence behavior, different extrapolations to the limit of the one-electron basis have been proposed and these are discussed in the current Section. The second approach is based on the incorporation to the Slater determinant of additional basis functions which depend explicitly on the coordinates of two electrons. The modern formulation of the explicitly-correlated theory at the level of second-order Møller-Plesset perturbation theory (MP2) will be briefly discussed in Section 2.3.

2.1.1 Partial-wave expansion

The ground-state helium wave function may be written as a product of a symmetric spatial part and an antisymmetric spin part

$$\Psi(\mathbf{x_1}, \mathbf{x_2}) = \Psi(\mathbf{r}_1, \mathbf{r}_2)2^{-1/2}[\alpha(1)\beta(2) - \beta(1)\alpha(2)] \,, \tag{2.1}$$

where $\mathbf{r}_i$ is the position vector of electron i and $\mathbf{x}_i$ the combined spatial and spin coordinates. In terms of a complete set of AOs $\phi_p(\mathbf{r})$, the spatial part may be expanded as [14]:

$$\Psi(\mathbf{r}_1, \mathbf{r}_2) = \sum_{p \geq q} c_{pq}[\phi_p(\mathbf{r}_1)\phi_q(\mathbf{r}_2) + \phi_q(\mathbf{r}_1)\phi_p(\mathbf{r}_2)] \,. \tag{2.2}$$

In terms of such AOs, the spatial part of the ground-state helium wave function in Eq. (2.1) may be decomposed into partial waves

$$\Psi(r_1, r_2, \theta_{12}) = \sum_{\ell=0}^{\infty} \Psi_\ell(r_1, r_2)P_\ell(\cos\theta_{12}) \,. \tag{2.3}$$

where the two-electron angular part $P_\ell(\cos\theta_{12})$ is the Legendre polynomial of degree ℓ in the cosine of the angle θ_{12} between the position vectors of the two electrons. In other words, adding a complete set of basis functions with angular momentum ℓ, with principal quantum number

$n = \ell + 1, \ell + 2, \ldots, \infty$, and $m_\ell = -\ell, \ldots, \ell$, introduces an energy expansion which leads to the exact electronic energy, as ℓ approaches infinity.

The error in energy obtained for the ground-state of the helium atom from a truncated partial wave expansion (for example, by omitting all terms $\ell > \mathcal{L}$), converges slowly with increasing $\mathcal{L}$. This means that a sufficiently large number of partial-waves should be introduced in the expansion of Eq. (2.3). Nevertheless, knowledge obtained from these expansions has been used either for the formulation of extrapolation schemes based on the asymptotic form, or to the employment of a more fundamental ansatz, which contains explicitly r_{12} terms. A brief presentation of these extrapolation schemes will be given in the next paragraphs.

In the early 1960s, Schwartz, based on the partial-wave expansion, proposed an asymptotic formula for the second-order energy of the $1/Z$ perturbation expansion of the ground state of two-electron atoms [15, 16]

$$\Delta E_\ell^{(2)} = -\frac{45}{256}(\ell + \frac{1}{2})^{-4} + \frac{225}{1024}(\ell + \frac{1}{2})^{-6} + \ldots . \tag{2.4}$$

Eq. (2.4) has been obtained from a perturbative treatment after partitioning the Hamiltonian of the helium atom into a zero-order bare-nucleus Hamiltonian $\hat{\mathcal{H}}_0 = -\frac{1}{2}\nabla_1^2 - \frac{1}{2}\nabla_2^2 - Z/r_1 - Z/r_2$ with nuclear charge Z and with $V = r_{12}^{-1}$ as the perturbation operator. In this study the basis sets were saturated at every angular momentum level ℓ. These partial-wave expansions for atoms and molecules include inverse powers of the highest angular momentum value present in the basis set (L). The slow convergence arises from the singularity in the perturbation operator r_{12}^{-1} at the electron coalescence.

A ℓ-dependent asymptotic formula was obtained by Hill [17] by means of the Rayleigh-Ritz variational method. The application of this theory to a configuration-interaction (CI) calculation on the ground state of the helium atom adds a contribution to the total energy from each partial wave of the form

$$E_\ell^{CI} = -3C_1(\ell + \frac{1}{2})^{-4} - 4C_2(\ell + \frac{1}{2})^{-5} + \ldots \tag{2.5}$$

where the coefficients $C_1, C_2, \ldots$ depend on the value of the electronic wave function at the points of coalescence. Eq. (2.5) differs from the perturbation-theory expansion of Eq. (2.4) in the fact that also odd-order terms contribute to the energy.

Kutzelnigg and Morgan [18] studied the partial-wave convergence in more detail. They found that for states of orbital momentum L the energy converges as $(\ell + 1/2)^{-4}$ for singlet states of parity $(-1)^L$ (natural-parity states), as $(\ell + 1/2)^{-6}$ for triplet states and as $(\ell + 1/2)^{-8}$ for singlet states of parity $(-1)^{L+1}$ (unnatural-parity states).

Two reasons make the application of the partial-wave formulae not a good choice for the correlation energy extrapolation of molecules. Firstly, the angular momentum is not a good quantum number for molecules and therefore, direct application to larger systems is not reliable. Secondly, the molecular AO basis sets are usually not constructed in such a manner that function spaces of a given (atomic) angular-momentum quantum number are saturated before the next function space is added. For example, the correlation-consistent (cc-pVXZ) family of basis sets [19–22] have been constructed in a way that convergence is obtained by adding functions of different angular momenta simultaneously, if they give similar energy lowerings. In other words, these systematic sequences of AO basis sets are based on the principal quantum number n. Expansions formulated by taking into account this principle are often called "principal expansions". [7]

2.1.2 Principal expansion

A more systematic way to improve atomic orbital basis sets for correlated calculations is the use of the principal quantum number n for defining systematic sequences of atomic orbital basis sets or for constructing natural orbital (NO) expansions. The radial form of the natural orbitals is determined by diagonalization of the one-electron density matrix in a large set of AOs. For helium, the NOs have the same nodal structure as the hydrogenic and Laguerre functions – in particular, the total number of nodes (radial and angular) is given by $n-1$, where n is the principal quantum number of the NO.

Carroll *et al.* [23] found that each NO provides to a good approximation an amount of energy to the ground-state helium atom proportional to

$$\epsilon_{n\ell m} \approx -a_\ell (n - \frac{1}{2})^{-6}, \tag{2.6}$$

with $a_0 \approx 0.24$ for $\ell = 0$ and $a_\ell \approx 0.21$ for $\ell > 0$. It becomes clear that for $\ell > 0$ and large n, the principal quantum number of the NO determines solely the energy contribution of each orbital. In addition, all n^2 orbitals belonging to the same shell simultaneously should always be included, noting that these all make the same contribution to the energy. In this principal expansion of the helium wave function, each new shell of NOs contribute an amount of energy proportional to n^{-4} to the ground state. Therefore, the principal expansion to the energy can be re-written as

$$\epsilon_n = \sum_{\ell=0}^{n-1} \sum_{m=-\ell}^{\ell} \epsilon_{n\ell m} = A_1 n^{-4} + A_2 n^{-5} + \dots \tag{2.7}$$

where A_i are constants independent of n. [24, 25] Noteworthy is that Eq. (2.7) has an asymptotic form in agreement with the partial-wave expansion of Eq. (2.5).

Approaches based on the principal expansion have been applied successfully to polyatomic systems [26–30], by means of increase of the cc-pVXZ family of basis sets, where $X = \{D, T, Q, 5, 6\}$ is the cardinal number of the basis set. When the cc-pVXZ basis sets are used, the highest angular momentum in the basis is $L = X - 1$ for H and He and $L = X$ for Li-Ar. In particular, Martin [26] used the formula

$$E_X = E_{\text{CBS}} + A(X + a)^{-4}, \tag{2.8}$$

where E_X is the CCSD(T) correlation energy for a specific correlation-consistent basis set and E_{CBS} the corresponding complete basis set (CBS) limit. The choice of $(X + a)$ for substituting L, where $a = \frac{1}{2}$, was made as a compromise between $a = 0$ (for hydrogen and helium) and $a = 1$ (for first- and second-row atoms). Martin [27] also proposed a formula with an additional $(X + \frac{1}{2})^{-6}$ term. Martin and Taylor [28] considered an additional formula with a variable exponent e where optimal values were found in the range $3.5 < e < 4.5$.

An alternative fit involving the inverse power of X was suggested by Helgaker *et al.* [30]

$$E_X = E_{\text{CBS}} + aX^{-3}. \tag{2.9}$$

In the original paper the basis set convergence of the water molecule was discussed in comparison with explicitly-correlated results. It was followed by a study on the Ne, N_2 and H_2O energies with quintuple and sextuple zeta quality basis sets. [31] The success of Eq. (2.9) was not only due to the excellent results but also due to the simplicity of the two-point fit for estimating the basis set limit. The simple analytical form

$$E_{XY}^{\text{corr}} = \frac{E_X^{\text{corr}} X^3 - E_Y^{\text{corr}} Y^3}{X^3 - Y^3} \tag{2.10}$$

can be used with the correlation energies from only two calculations with the cc-p(C)VXZ basis sets, where $Y = X - 1$.

By using different two-point fit extrapolations for each level of theory, Schwenke [32] proposed the formula

$$E_{\text{CBS}} = (E_{X+1} - E_X)F_{X+1}^C + E_X, \tag{2.11}$$

which involves empirical parameters (F_{X+1}^C) that are specific for the HF, CCSD and (T) contributions to the total energy. Despite the small set of molecules from which the parameters have been obtained (first row molecules), Eq. (2.11) is remarkably applicable beyond the first row molecules.

Bakowies [33] defined a trial function $X^{-\beta}$ which, for an effective exponent $\beta = \beta_{\text{eff}}(X, X + 1, X + N)$, provides the correct energy E_{X+N}, when results are extrapolated from two smaller basis sets E_X and E_{X+1}. For both MP2 and CCSD, the β_{eff} varies monotonically with the target of extrapolation. By testing this approach on a test-set of 105 molecules containing H, C, O, N and F, he obtained a single "optimal" exponent β_{opt}.

For the sake of completeness, exponential fits of the atomic correlation energy should be mentioned. In 1992, shortly after the introduction of the cc-pVXZ basis sets, Feller proposed a simple form

$$E_X = E_{\text{CBS}} + Ae^{-bX} \tag{2.12}$$

to estimate the binding energy of the water dimer at the basis set limit. [34] The three parameters E_{CBS}, A and b were fit to the sequence of three energies. Significantly more accurate results are obtained with the addition of a Gaussian term in Eq. (2.12), as it was suggested by Peterson et al.. [35] Feller et al. [36] recently discussed in detail the behavior of different extrapolation schemes of the CCSD(T) correlation energy. They concluded that almost all these two- or three-point fits achieve accuracy obtained normally from one higher order basis set. They also highlight that different basis sets are performing better with specific formulas, like, for example, a $X =$ D, T and Q sequence of basis sets provides superior results with the three-parameter, mixed Gaussian/exponential expression.

2.1.3 Natural orbital expansion

30 years ago, Petersson et al., based on a natural orbital expansion, showed that a higher-order energy in the CBS limit can be estimated from the second-order basis set truncation error. [13] In particular, they obtained an asymptotic formula for pair natural orbitals (PNOs) expansions based on perturbation theory. They proposed that the infinite-order energy can be computed from the first N natural orbitals (NOs) from the formula

$$E(\infty) \approx E(N) - \left(\sum_{\mu=1}^N C_\mu \right)^2 \frac{225}{4608}(N + \delta)^{-1}, \tag{2.13}$$

where, in order to fit the energies for small N, an empirical parameter ($\delta = 0.363$ for the helium atom) was introduced. The C_μ is the coefficient of the configuration-interaction (CI) expansion. Each configuration is described from the principal quantum number n and the angular momentum ℓ of a NO. All coefficients are negative, except those that correspond to $n = 1$ and $\ell = 0$. Therefore, their squared sum is always less than one and represents an *interference effect* of the higher NOs of the multi-configuration wave function which reduces the correlation energy of the first NO. [37] Based on this, Petersson and co-workers formulated the CBS family of model chemistries. [38] In these model chemistries the interference factors are computed from the coefficients of the first-order wave function. The second-order interference effect can be used to extrapolate the infinite-order pair energies calculated from a finite number of N pair natural orbitals. The extrapolation is different for every electron pair; an interference factor is being calculated from the second-order amplitudes for every same-spin and opposite-spin pair.

In the next section, the theoretical background of this extrapolation will be presented as far as the interference factor constitutes not only the core of the CBS extrapolation, but also of the present Thesis. Therefore, the derivation of the interference factor, the use of it within the framework of the CBS model chemistries plus some technical considerations on its optimum application will be discussed.

2.2 CBS Extrapolation and Model Chemistries

2.2.1 Interference effects in pair correlation energies

In 1981, Marc Nyden and George Petersson [13] used a generalization of the overlap approximation for the generalized valence bond (GVB) pair correlation energy [1]. Based on that, they derived an expression for the correlation energy of the ground state of the helium atom in terms of perturbation theory in a general multi-configuration self-consistent-field (MCSCF) wave function. It should be reminded that after a proper transformation and for the same set of basis functions, the GVB wave function can be written in an identical form with the MCSCF wave function. [39] The asymptotic form of the convergence of a second-order intraorbital pair correlation energy $e_{ii}^{(2)}$ in an N_{ii}-configuration pair natural orbital (PNO) expansion is (in a.u.)

$$\delta e_{ii}^{(2)}(N_{ii}) = (-225/4608)(N_{ii} + \delta_{ii})^{-1}, \tag{2.14}$$

where $\delta e_{ii}^{(2)}(N_{ii})$ is the basis set truncation error after the first N_{ii} (PNO) configurations for pair ii, and δ_{ii} is a constant associated with pair ii, important for fitting the energies for a small N. In other words, the right-hand side of Eq. (2.14) provides a second-order pair energy correcting term at the limit of the infinite (complete) expansion of PNOs. The asymptotic form for the convergence of an infinite-order (i.e. pair CI) pair correlation energy e_{ii} is obtained from

$$\delta e_{ii}(N_{ii}) = \left(\sum_{\mu=1}^{N_{ii}} C_\mu \right)^2 (-225/4608)(N_{ii} + \delta_{ii})^{-1}, \tag{2.15}$$

---

[1]For sake of simplicity, the terms "pair energy" or "pair correlation energy" correspond to "pair *electron* correlation energy".

where C_μ is the coefficient of configuration μ in the multi-configuration wave function obtained from the NO expansion of the singlet ground state

$$^1\Psi_{N_{ii}}(1,2) = {}^1\Theta(1,2)\sum_{\mu=1}^{N_{ii}}C_\mu\varphi_\mu(1)\varphi_\mu(2) \tag{2.16}$$

$$^1\Theta(1,2) = \frac{1}{\sqrt{2}}\{\alpha(1)\beta(2) - \beta(1)\alpha(2)\}, \tag{2.17}$$

where φ is a NO and $^1\Theta(1,2)$ the usual two-electron singlet spin function. This estimation of the truncation error of the correlation energy is based on Schwartz's asymptotic formula (Eq. (2.4)) for the contribution to the second-order energy. However, N_{ii} should correspond to a closed shell in order that Eq. (2.15) provides accurate results. This means that shell functions should be filled accordingly to a principal number n. For example, $N_{ii} = 1$ for 1s, $N_{ii} = 5$ for 1s2s2p, $N_{ii} = 14$ for 1s2s2p3s3p3d shells of NOs, and so on. A more detailed explanation on the derivation of Eq. (2.15) is given in Appendix B.1. The same authors had also proven [37] that their approach is identical with the principal expansion of Carroll et al. [23].

The asymptotic form of the PNO (or MCSCF) correlation energy is a product of two factors, the squared sum of the CI coefficients and the pair correlation energy. If the MCSCF calculation includes all configurations which are occupied in a closed shell system, or the complete expansion of the PNOs, the latter term is expected to be constant. Therefore, the correlation energy will be dominated by changes of the $\left(\sum_{\mu=1}^{N}C_\mu\right)^2$ term. This can be interpreted as emphasizing the importance of an *interference effect* from the coefficients of the PNOs, giving in that manner the appropriate weight to the corresponding pair energy. This *interference effect* relates the pair energies $^{\alpha\beta}e_{ij}$ and $^{\alpha\alpha}e_{ij}$ from the second-order perturbation theory to the infinite-order CI theory. Each extrapolation for a single pair energy is scaled by an individual interference factor $\left(\left(\sum_{\mu=1}^{N}C_\mu\right)^2\right)$. [40] This is of high importance, considering that every pair energy shows a different asymptotic convergence.

In principle, the interference factor is constructed from the GVB (or MCSCF) CI coefficients. A main feature of the GVB approach is the fact that it properly describes the dissociation of molecules in open shell fragments. This is a result of the way that the coefficients are constructed. For example, for a two-electron singlet state, the GVB orbitals can be constructed from the first two natural orbitals:

$$\phi_1 = \frac{\phi_\alpha + \phi_\beta}{\sqrt{2(1+S_{\alpha\beta})}}, \qquad \phi_2 = \frac{\phi_\alpha - \phi_\beta}{\sqrt{2(1-S_{\alpha\beta})}} \tag{2.18}$$

with the coefficients of the natural orbital configurations

$$C_1 = \frac{1+S_{\alpha\beta}}{\sqrt{2(1+S_{\alpha\beta}^2)}}, \qquad C_2 = -\frac{1-S_{\alpha\beta}}{\sqrt{2(1+S_{\alpha\beta}^2)}} \tag{2.19}$$

The interference factor retains this property and leads to a smooth extrapolation to the limit of the basis set truncation error.

2.2.2 Asymptotic extrapolations to the Complete Basis Set limit

Extension of the above for a generalization to a multi-electron system needs another important consideration. Eqs. (2.14) and (2.15) should be modified separately for occupied α and β spin-orbitals ($\alpha\beta$-pairs) and $\sigma\sigma$ same-spin pairs, either $\alpha\alpha$ or $\beta\beta$. Therefore, the intraorbital pair energies e_{ii} and the $\alpha\beta$- and $\sigma\sigma$-interorbital pair energies ($^{\alpha\beta}e_{ij}$ and $^{\sigma\sigma}e_{ij}$) should be considered separately since each type of pair interaction shows a different asymptotic convergence as the basis set increases. A brief description of this task is given in Appendix B.2.

Based on Eqs. (2.14) and (2.15), extrapolations to the complete basis set of the second- and infinite-order correlation energy. Firstly, the MP2 limit will be discussed. In the original work of Petersson, this is estimated from the extrapolation formulas

$$^{\alpha\beta}e_{ij}^{(2)}(N) = {}^{\alpha\beta}e_{ij}^{(2)}(\text{CBS}) + {}^{\alpha\beta}f_{ij}\frac{225}{4608}(N + {}^{\alpha\beta}\delta_{ij})^{-1} \tag{2.20a}$$

$$^{\sigma\sigma}e_{ij}^{(2)}(N) = {}^{\sigma\sigma}e_{ij}^{(2)}(\text{CBS}) + {}^{\sigma\sigma}f_{ij}\frac{225}{4608}(N + {}^{\sigma\sigma}\delta_{ij})^{-5/3}. \tag{2.20b}$$

The expressions for the overlap factors $^{\alpha\beta}f_{ij}$ and $^{\sigma\sigma}f_{ij}$ are given in Appendix B.2. The $\alpha\beta$- and $\sigma\sigma$-pairs of the above equations do not represent spin-adapted singlet and triplet pairs; rather, the $\alpha\beta$-pair contributes to both singlet and triplet pairs.

In Eqs. (2.20a) and (2.20b), $e_{ij}^{(2)}(\text{CBS})$ and δ_{ij} constitute the fitting parameters of the extrapolation and are obtained from two-point fits. The first point corresponds to the HF calculation; $N = 1$ and thus, $e_{ij}^{(2)}(1) = 0$. A selected range on different N values is then tried for the second point, obtaining every time the corresponding pair energy $e_{ij}^{(2)}(N)$. For each N, the two equations are solved for the two unknowns and the most negative $e_{ij}^{(2)}(\text{CBS})$ is taken as the final, extrapolated second-order pair energy.

As a next step, the interference effect is used for the extrapolation of the estimated second-order $e_{ij}^{(2)}(\text{CBS})$ energies to the infinite-order.

$$e_{ij}^{(\infty)}(\text{CBS}) = e_{ij}^{(\infty)}(N) + \left(\sum_{\mu=1}^{N} C_\mu\right)^2 [e_{ij}^{(2)}(\text{CBS}) - e_{ij}^{(2)}(N)], \tag{2.21}$$

which is a generalization of Eq. (2.15), for the helium atom. The interference factor takes values between zero and one and reduces significantly the truncation errors in the infinite-order pair energies. It can also be applied as a fine recipe for the estimation of errors of higher level methods from the corresponding lower-level MP2 errors. [7, 40, 41] Eq. (2.21) can be rewritten as

$$\delta e_{ij}^{(\infty)} = \left(\sum_{\mu=1}^{N} C_\mu\right)^2 \delta e_{ij}^{(2)}, \tag{2.22}$$

where

$$\delta e_{ij}^{(2)} = e_{ij}^{(2)}(\text{CBS}) - e_{ij}^{(2)}(N) \tag{2.23}$$

is the second-order basis set truncation error and

$$\delta e_{ij}^{(\infty)} = e_{ij}^{(\infty)}(\text{CBS}) - e_{ij}^{(\infty)}(N) \tag{2.24}$$

the higher-order basis set truncation error.

2.2.3   The CBS family of model chemistries

A "theoretical model chemistry" is a complete algorithm for the calculation of the energy of any geometry of any molecular system. By definition, it should fulfill two features. Firstly, it should not involve subjective decisions in its application from the user and secondly, it should be size consistent in order that the energy of every molecular species is defined uniquely. Satisfaction of the variational principle or invariance to unitary transformations among orbitals are desirable criteria but not required.

The definition of a model chemistry was given by Pople [42] and based on it, Petersson and co-workers [38, 43–49] formulated a family of model chemistries under the name "CBS". Central point of their composite schemes are the above mentioned asymptotic extrapolations. The CBS model chemistry requires a set of basis functions for every level of theory included, a method obtaining the self-consistent-field (SCF) CBS energy and a method for obtaining the CBS correlation energy. The basis set of choice consists of atomic pair natural orbitals (APNO). [38, 43] The APNO expansion of the SCF orbitals is more rapidly convergent than the conventional contracted Gaussian type orbital expansions. They also affirm an extrapolation of pair correlation energies from small PNO calculations to estimate the CBS values. However, the APNOs are indeed a really large basis set: (14s9p4d2f,6s3p1d)/[6s6p3d2f,4s2p1d]. For that reason, different schemes have been proposed containing more compact basis sets. [44] Another, more compact APNO variant (named DZ+P{KK,KL,LL,LL'}) is also used in specific steps of the CBS composite scheme. The correlation energy is obtained from the extrapolation to the basis set limit second-order term and the interference corrected higher-order energy term. As a method of choice for higher-order terms, the quadratic CI (QCI) [50] or the coupled-cluster theory is used.

From the CBS family, the most accurate and robust model chemistry is the CBS-QCI/APNO [45]. It includes a frozen-core (fc) QCI singles-and-doubles and perturbative triples (QCISD(T)) calculation with the (14s9p4d2f,6s3p1d)/[6s6p3d2f,4s2p1d] APNO basis set on experimental geometries or from QCISD(T)/6-311G** geometry optimizations. The basis set limit of the SCF is obtained from an l^{-6} extrapolation. [51] The difference between all-electron and fc-QCISD(T)/DZ+P{KK,KL,LL,LL'} calculations provides a core-correction contribution to the total energy. The second-order basis set limit is obtained from extrapolation of the fc-MP2 energy, using the Eqs. (2.20a) and (2.20b) with the large APNO basis set. The interference effect is used for the correction to the CBS extrapolation (Eq. (2.21)). An important difference from the previous subsection is that in the CBS model chemistries the interference factor is calculated from the first-order amplitudes of MP2 (*vide infra*). A third-order correction is added based on the extrapolation formula

$$\sum_{kl \neq ij} e_{ij,kl}^{(3)}(\text{CBS}) = \sum_{kl \neq ij} e_{ij,kl}^{(3)}(N_{ij}, N_{kl}) \left[\frac{e_{ij}(\text{CBS}) - e_{ij}^{(2)}(\text{CBS})}{e_{ij}(N_{ij}) - e_{ij}^{(2)}(N_{ij})} \right], \qquad (2.25)$$

where the pair couplings are scaled by the same fraction as the high-order contributions to the associated pairs. Zero-point energy corrections are also added from vibrational frequencies calculated at the unrestricted HF (UHF)/6-311G** level of theory. Finally, an empirical term is included in this model chemistry, based on the overlap of α and β spin orbitals. This empirical correction varies smoothly as the bond dissociates to the proper reduced value for the separate atoms

$$\Delta E_{\text{emp}} = -0.00174 \sum_{i} \left(\sum_{\mu=1}^{N_{ii}} C_{\mu ii} \right)^2 |S|_{ii}^2, \qquad (2.26)$$

where $(\sum C_{\mu ii})^2$ is the square of the trace of the first-order wave function and $|S|_{ii}$ the absolute overlap integral (Appendix (B.2)). In later CBS versions, like the CBS-Q or CBS-4 models, a correction for the spin contamination of the unrestricted cases was added. [46]

Among the various members of the CBS family of model chemistries, the CBS-Q model [46] deserves some further discussion because it provides a reasonable compromise between the high accuracy of the CBS-QCI/APNO model and the computational speed of more compact models like, for example, the CBS-4 model. The most important differences from CBS-QCI/APNO are mainly three. Firstly, the geometry optimization step, which is done at the fc-MP2/6-31G† level of theory. In the more recent CBS-Q variety, named CBS-QB3 [47], this step has been substituted by a density functional theory (DFT) optimization with the B3LYP functional. Secondly, the core correlation correction does not use a CBS extrapolation at the second-order with the DZ+P{KK,KL,LL,LL$'$} basis set but it is obtained from an empirical formula based on CBS calculations on the Na$^+$, Na and Na$^-$ species. Another important difference is the reasonable reduction of the basis sets; instead of the large APNO, the more compact 6-311G($2df$)/6-311G($3d2f$) basis sets are used. In 2006, Wood et al. [49] proposed a restricted-open shell CBS model chemistry (ROCBS-QB3), in which QCISD(T) has been replaced from the unrestricted CCSD(T).

2.2.4 Other model chemistries

Apart from the CBS model chemistries, a variety of different complete protocols have been reported in the literature. Historically, the first such model chemistry was introduced by Pople and co-workers in 1989, named Gaussian-1 (G1). [52] G1 attempted to approximate a high-level but computational demanding calculation [QCISD(T)/6-311+G(2df,p)] with lower cost and in a "black-box" manner. It used an additivity scheme with different levels of theory and different basis sets for each level, similar with the CBS theories described above. Improved accuracy was achieved from the updated G2 theory. [53] It started with the G1 energy and added a second-order correction in order to account with an error arising from the additivity assumption of G1. It also introduced an empirical, "higher-level correction", based on the number of electron pairs. In 1998, a new modified Gn theory appeared (G3), which replaced various steps of G2, incorporated a new, but still empirical, higher-level correction, as well as corrections for spin-orbit effects and core/valence correlation. [54] G3 achieved to reduce the mean-absolute deviation (MAD) of a test set composed of 148 experimental enthalpies of formation at 298 K from 1.56 (G2) to 0.94 kcal/mol. The last member of the Gn models (G4) was published in 2007, which included separate HF extrapolation to the basis set limit, additional d functions to the basis set, CCSD(T) theory instead of QCISD(T) and two new higher-level corrections instead of the previous. [55]

Similar in philosophy are the Weizmann-n (Wn) model chemistries by Martin and co-workers. W1 and W2 include separate CBS extrapolations for SCF, CCSD and (T) energy components and they differ on the size of the basis sets, with the W2 being more expensive. [56] Five years later, a major W3 revision was published. [57] W3 determines the contribution of triples excitations from CCSDT calculations with the cc-pVDZ and cc-pVTZ basis sets. For a test set of 30 small molecules, W3 provided a smaller unsigned deviation than W2 (0.22 instead of 0.40 kcal/mol). W4 [58] is using CCSD(T)/cc-pVQZ reference geometries and yields a MAD of 0.15 kcal/mol for the atomization of 26 small molecules. The applicability of the Wn model chemistries was recently extended with the addition of explicitly-correlated variants. [59] Basis sets with one cardinal number less were used, without loss of accuracy, in comparison with the conventional Wn models.

In 2004, Tajti et al. introduced the high-accuracy extrapolated ab initio thermochemistry (HEAT) model which was targeting for highly accurate enthalpies of formation of atoms and small molecules without any dependence to empirical scale factors. [60] The fc-CCSD(T) basis set limit was estimated from an extrapolation of aug-cc-pCVQZ and aug-cc-pCV5Z energies, based on Eq. (2.9). Extrapolated frozen-core high-order corrections from CCSDT/cc-pV(TQ)Z and CCSDTQ/cc-pVDZ calculations were added and anharmonic zero-point vibrational energies were evaluated at the fc-CCSD(T)/cc-pVQZ level of theory. For a test set of 26 small molecules, HEAT produced a mean unsigned deviation of 0.09 kcal/mol. In different variants of the HEAT protocol, more accurate, three-point extrapolations have been explored and a diagonal Born-Oppenheimer correction has been added. [61, 62]

An alternative family of models to the Gn model chemistries was proposed by DeYonker et al., based on a composite method but without relying on empirical corrections. [63] The "correlation-consistent composite approach" (ccCA) is faster than the Wn theories and like the HEAT protocol, it includes corrections to the correlation energy obtained from different basis sets. The ccCA-CBS-1 and ccCA-CBS-2 implementations are of particular interest for the current study, because they include extrapolations to the basis set limit based either on an exponential (Eq. (2.12)) [34] or a mixed Gaussian/exponential formula [35]. Even if the corresponding MAD for specific sets of molecules exceeded the so-called chemical accuracy (1 kcal/mol), the ccCA model chemistries have been used successfully to relative large molecules with transition metals [64, 65] or in molecules with a multi-reference character [66].

The accuracy that these model chemistries can achieve is given by comparing statistical results between them. An illustrative example is the Gaussian-2 (G2) test set, which includes experimental energies of 125 molecules. [53] The MAD of the CBS-4, CBS-Q and CBS-QCI/APNO from experimental data of the G2 test set are 2.0, 1.0 and 0.5 kcal/mol, respectively. [46] The G2 model chemistry of Pople [53] has a MAD of 1.2 kcal/mol. A more recent study of Martin [58] on the performance of various DFT functionals and model chemistries (composite schemes) for thermochemistry data provides some more comparative results. The different functionals and composite schemes were tested for the calculation of the atomization energies of 140 molecules. All model chemistries have a significant lower MAD than the DFT functionals. CBS-QB3 and ROCBS-QB3 have comparable MAD (1.5 and 1.4 kcal/mol, respectively), with G4 theory [55] having a MAD of 0.9 kcal/mol and the various unrestricted flavors of the Weizmann-1 (W1) [67] models a MAD of 0.6 kcal/mol. Another computational study [68] on the hydrogen abstraction from small alkanes showed that the CBS-APNO method should be used for high accuracy. These results were calibrated with the linear CCSD(T)-R12 method.

Peterson et al. [69] recently concluded that for any model chemistry or composite approach, fc-CCSD(T) atomization energies which target better than 1 kcal/mol of accuracy should either be carried out with a basis set of at least sextuple-zeta quality or an extrapolation between aug-cc-pVQZ and aug-cc-pV5Z basis sets. These requirements may differ by using explicitly-correlated coupled-cluster methods. In particular, the same authors concluded that CCSD(T)-F12b/cc-pVQZ-F12 or extrapolated values from CCSD(T)-F12b/cc-pV(TQ)Z-F12 calculations can obtain an accuracy in the 1-2 kJ/mol range.

2.3   Explicitly-Correlated MP2

Cornerstone concept of the present study is the substitution of the two-fit second-order extrapolation to the CBS pair energies ($e_{ij}^{(2)}$(CBS) of Eq. (2.21)) with modern quantum chemistry methods that take care *explicitly* the electron correlation in atoms and molecules. These "explicitly-

Table 2.1: A summary of functions and index conventions.

$\phi_i, \phi_j, \ldots$	Spin orbitals
$\varphi_i, \varphi_j, \ldots$	Spatial functions
$\sigma, \tau, \ldots$	Spin functions
$p, q, r, \ldots$	Orbitals in the HF basis
$i, j, k, \ldots$	Active occupied orbitals
$a, b, c, \ldots$	Active virtual orbitals in the finite basis
$\alpha, \beta, \gamma, \ldots$	Virtual orbitals in a formally complete basis
$v, w, x, \ldots$	Orbitals of the F12 geminal basis

correlated" or "R12/F12" methods have as main advantage the fact that they can calculate electronic correlation energies of molecular ground and excited states close to the limit of a complete basis set.

In the present section, a brief introduction to the Møller-Plesset second-order perturbation theory [70] will be given. MP2 theory is central to this study because the interference factor is calculated from its first-order amplitudes. As a next step, the principles and modern aspects of the explicitly-correlated second-order perturbation theory will be discussed.

2.3.1 MP2

All equations are presented in the formalism of second-quantization in order to keep a consistency with the literature. Table 2.1 includes all the index notation, which will also be followed in the next chapters. As reference wave function the HF determinant is considered in all cases. The antisymmetric product of spin orbitals represents a state where each electron behaves as an independent particle, which is however subject to Fermi correlation. [71] This observation suggests that the optimal determinant, which is the $|\mathrm{HF}\rangle$ determinant, is obtained by solving a set of effective one-electron Schrödinger equations for the spin orbitals. This is feasible by using as Hamiltonian the *Fock operator*

$$\hat{f} = \sum_{pq} f_{pq} a_p^\dagger a_q \tag{2.27}$$

for solving the Hartree-Fock equations (or simpler Fock equations), where the elements f_{pq} constitute the *Fock matrix*. The Fock equations are solved by diagonalization of the Fock matrix. The resulting eigenvectors are called the *canonical spin orbitals* of the system and the orbital energies are the eigenvalues of the Fock matrix. The iterative procedure, where the Fock matrix is repeatedly reconstructed and rediagonalized until the spin orbitals generated by its diagonalization become identical to those from which the Fock matrix has been constructed, is called self-consistent-field (SCF) method. [72] For a given electronic state, the electron correlating effects contribute to the total nonrelativistic energy (i.e. the full CI (FCI) energy) of the electronic system. The *correlation energy* is defined as

$$\delta E_{\mathrm{corr}} = E_{\mathrm{exact}} - E_{\mathrm{HF}} \tag{2.28}$$

The correlation energy [2] is described in terms of a complete one-electron basis, as it is discussed in Section 2.1. In practice, however, an incomplete basis must be used.

---

[2]Throughout this Thesis, the correlation energy component of the total electronic energy will be written as δE. The total energy of one method, i.e. MP2, will be given as $E = E_{\mathrm{HF}} + \delta E_{\mathrm{MP2}}$.

In Møller-Plesset perturbation theory, the electronic Hamiltonian is partitioned as

$$\hat{\mathcal{H}} = \hat{f} + \hat{\Phi} + h_{\text{nuc}} \tag{2.29}$$

with $\hat{\Phi}$ the *fluctuation potential* and h_{nuc} the nuclear-nuclear term. The fluctuation potential represents the difference between the true two-electron Coulomb potential $\hat{g}$ of the spin-free non-relativistic electronic Hamiltonian operator and the effective one-electron Fock potential of the Fock operator.

In MP2 theory, the Fock operator represents the zero-order operator and the fluctuation potential the perturbation. The zero-order electronic state is represented by the HF state in the canonical representation and the corresponding zero- and first-order energy terms are

$$E^{(0)} = \langle \text{HF}|\hat{f}|\text{HF}\rangle = \sum_i \epsilon_i \tag{2.30}$$

$$E^{(1)} = \langle \text{HF}|\hat{\Phi}|\text{HF}\rangle \tag{2.31}$$

where the summation is over the occupied spin orbitals. Thus, the Hartree-Fock energy is equal to the sum

$$E_{\text{HF}} = E^{(0)} + E^{(1)} + h_{\text{nuc}} = \langle \text{HF}|\hat{\mathcal{H}}|\text{HF}\rangle \tag{2.32}$$

In the spin-orbital formalism, the singly and doubly excited determinants can be written as

$$|\Psi_i^a\rangle = a_a^\dagger a_i|\text{HF}\rangle \tag{2.33a}$$

$$|\Psi_{ij}^{ab}\rangle = a_a^\dagger a_i a_b^\dagger a_j|\text{HF}\rangle \tag{2.33b}$$

$$|\Psi_{ijk}^{abc}\rangle = a_a^\dagger a_i a_b^\dagger a_j a_c^\dagger a_k|\text{HF}\rangle \, . \tag{2.33c}$$

The introduction of the cluster operators will simplify the notation of the following equations. The cluster operators for single and double excitations are

$$\hat{T}_1 = \sum_{ia} t_a^i a_a^\dagger a_i \tag{2.34a}$$

$$\hat{T}_2 = \sum_{\substack{i>j \\ a>b}} t_{ab}^{ij} a_a^\dagger a_i a_b^\dagger a_j \, , \tag{2.34b}$$

where t_a^i and t_{ab}^{ij} are the corresponding amplitudes.

Before proceeding to the MP2 energy, a few considerations should be taken into account. According to the Brillouin theorem, the Hamiltonian does not couple the HF wave function with singly excited determinants. [14] Also, since the Hamiltonian is a two-electron operator, it does not couple determinants that differ by more than two levels of excitation. Therefore, only doubly excited configurations interact directly with the HF state through the Hamiltonian operator.

The first-order wave function $|\Psi^{(1)}\rangle$, from which the MP2 energy will be obtained, can be written as

$$|\Psi^{(1)}\rangle = \hat{T}_2^{(1)}|\text{HF}\rangle = \sum_{\substack{i>j \\ a>b}} t_{ab}^{ij(1)} a_a^\dagger a_i a_b^\dagger a_j|\text{HF}\rangle \, , \tag{2.35}$$

where the superscript (1) stands for the first-order components. The double excitation amplitudes $t_{ab}^{ij(1)}$ are determined by requiring the projection of the first-order Schrödinger equation on the doubly excited determinants (Eq. (2.33b)). Thus, the MP2 correlation energy is given by

$$\delta E_{\mathrm{MP2}} = \langle \mathrm{HF} | \hat{\Phi} | \Psi^{(1)} \rangle = \sum_{\substack{i>j \\ a>b}} t_{ab}^{ij} g_{ij}^{ab}, \tag{2.36}$$

where g_{ij}^{ab} are the matrix elements

$$g_{ij}^{ab} = (ia|jb) - (ib|ja) \tag{2.37}$$

which are composed from the $(ia|jb)$ two electron integrals. These can be written in Mulliken notation as

$$(ia|jb) = \delta_{\sigma\sigma'}\delta_{\tau\tau'} \int \int \phi_i^{\sigma}(\mathbf{x}_1)\phi_j^{\tau}(\mathbf{x}_2)r_{12}^{-1}\phi_a^{\sigma'}(\mathbf{x}_1)\phi_b^{\tau'}(\mathbf{x}_2)\mathrm{d}\mathbf{x}_1\mathrm{d}\mathbf{x}_2. \tag{2.38}$$

The first-order t_{ab}^{ij} amplitudes are

$$t_{ab}^{ij} = \sum_{\substack{i>j \\ a>b}} \frac{g_{ij}^{ab}}{\epsilon_i + \epsilon_j - \epsilon_a - \epsilon_b}, \tag{2.39}$$

2.3.2 MP2-F12

As it has already been mentioned in the beginning of this chapter, an alternative approach to reach a faster convergence to the basis set limit is provided by the explicitly-correlated methods. Conventional methods like MP2 or CCSD use antisymmetrized products of one-particle orbitals (Slater determinants) to construct two electron and higher-rank basis sets. However, the exact wave function at short interelectronic distances is not described correctly from the Slater determinants. Based on the *a priori* knowledge of the exact wave function, alternative accurate trial wave functions can be constructed. In particular, a Slater determinant takes into account the fermionic nature of the electrons but it is the Coulombic nature of the interactions between quantum particles that defines the wave function at short interparticle separations. In order to be more specific, sharp features appear at the electron-nucleus and electron-electron coalescence points. [6]

In 1957, Kato showed that all eigenfunctions of the nonrelativistic Hamiltonian are continuous throughout configuration space and have bounded continuous first derivatives, except at the Coulomb singularities. [73] He also showed that such discontinuity at the coalescence point of two electrons can be expressed as

$$\left.\widetilde{\frac{\partial\psi}{\partial r_{12}}}\right|_{r_{12}=0} = \gamma\psi(r_{12}=0) \tag{2.40}$$

with $\gamma = \frac{1}{2}$. Similar conditions hold for the Coulomb singularity at the electron-nucleus coalescence. In this case, γ takes a value of minus the nuclear charge. The tilde indicates an average over a sphere around the singularity.

This cusp condition is obtained by considering the behavior near the coalescence point of electrons 1 and 2. The exact solution can be expressed in terms of r_{12} as a spherical harmonic $Y_{\ell m}$, with

ℓ determined by particle exchange and/or other symmetries, times the radial part with the r_{12}^{ℓ} leading term

$$\Psi(\mathbf{r_1}, \mathbf{r_2}, \ldots) \approx r_{12}^{\ell} \sum_{m=-\ell}^{\ell} \left(1 + \frac{r_{12}}{2(\ell+1)} + \mathcal{O}(r_{12}^2)\right) Y_{\ell m}(\theta_{12}, \phi_{12}) \Phi_m(\mathbf{R}_{12}, \ldots) \qquad (2.41)$$

In the absence of particle exchange or other symmetries, $\ell = 0$ and this condition

$$\Psi(\mathbf{r_1}, \mathbf{r_2}, \ldots) \approx \left(1 + \frac{1}{2} r_{12} + \mathcal{O}(r_{12}^2)\right) \Phi_0(\mathbf{R}_{12}, \ldots) \qquad (2.42)$$

is equivalent with Kato's result. This clarifies the cusp condition; the derivative of the wave function with respect to r_{12} is the same, independently of the direction in which the cusp is approached. Hence, crossing the coalescence point changes the sign of the first derivative. On the contrary, the first derivative at the coalescence point is continuous for the triplet case because the first-order spherical harmonic is an odd function.

Pack and Byers Brown reached the above approach in order to avoid the spherical averaging around the singularity and expressed Kato's cusp condition in a form including $\ell = 0$ and $\ell = 1$. [74] Kutzelnigg and Morgan were the first to reformulate the cusp condition for unnatural parity singlet states. [18] In 2008, Tew followed this approach to derive higher-order terms in the partial-wave expansion around the cusp. He concluded that the leading-order cusp condition is universal, while a system and state dependent term arises in the higher order conditions. [75]

It was long before Kato's study in the cusp conditions that the interelectronic limits were used for constructing efficient wave functions. Slater was the first who attempted to construct a wave function for a two-electron atom that satisfied both the Rydberg limit, in which one electron is very far from the nucleus, and the core limit, where both electrons are close to the nucleus. An even better approximation is acquired by also including the r_{12} coordinate. [76] However, the first successful calculation using wave functions including explicitly the interelectronic distance r_{12} was the calculation of the He ground state by Hylleraas in 1929 [77], by using the coordinates

$$s = r_1 + r_2; \qquad t = r_1 - r_2; \qquad u = r_{12}. \qquad (2.43)$$

In the Hylleraas expansion the spatial part of the He ground-state wave function is written as

$$\Psi_N = \exp(-\zeta s) \sum_{k=1}^{N} c_k s^{l_k} t^{2m_k} u^{n_k}. \qquad (2.44)$$

Only even powers of t contribute to singlet states, which have symmetric spatial part and an antisymmetric spin function. With only three terms, that is, with the spatial wave function

$$\Psi_3 = \exp(-\zeta s)(c_1 + c_2 u + c_3 t^2), \qquad (2.45)$$

Hylleraas obtained the energy $E = -2.90243 E_{\rm h}$, after variationally optimizing both the linear parameters c_k and the nonlinear parameter ζ. Hence, both Slater and Hylleraas are considered as the persons who introduced explicitly-correlated wave functions to electronic structure theory.

These two aspects, i.e. the introduction of the interelectronic distance r_{12} to the wave function and the appropriate account for the cusps of the electronic wave function, are essential for rapid decay of the basis set errors of many-electron wave functions. [78] Practical application of these ideas was halted due to major computational problems; the complexity and high dimensionality of the numerous integrals that need to be evaluated and computed made the utilization of the theory almost impossible for n-electron systems.

However, there was a tremendous progress the last years in this field, with the development of efficient methods for the calculation of the integrals. From these developments, two were the key for the advance of explicitly-correlated methods. The first was the weak orthogonality functional, introduced by Szalewicz et al. [79, 80] and modified by Tew et al. [81] for modern F12 methods, which reduces the dimensionality of the required integrals. The second and definitely most crucial development, was the groundbreaking idea of Kutzelnigg to factorize all the difficult many-electron integrals into one- and two-electron integrals by inserting the resolution-of-the-identity (RI) and to use a partial-wave analysis for its derivation. [82] The generalization of this idea towards molecular many-electron systems was firstly accomplished at the MP2 level. [83] In the original implementation, the conventional excitations were augmented by products of occupied one-electron orbitals by excitations of electron pairs into explicitly-correlated pair functions, which are constructed as $\hat{Q}_{12}r_{12}\phi_i\phi_j$. $\hat{Q}_{12}$ is the strong-orthogonality projector which retains the separability of the MP2 expressions into decoupled pair equations:

$$\hat{Q}_{12} = (1 - \hat{O}_1)(1 - \hat{O}_2); \qquad \hat{O}_1 = \sum_i |\phi_i(1)\rangle\langle\phi_i(1)|. \tag{2.46}$$

In modern explicitly-correlated methods, the linear r_{12} term has been substituted by a Slater-type function [84]

$$f_{12} = \frac{1}{\gamma}(1 - \exp(-\gamma r_{12})), \tag{2.47}$$

with a length-scale parameter γ. Due to this exponential ansatz, modern explicitly-correlated methods are called F12, while the R12 notation is hold for the linear expression $f_{12} = r_{12}$. The R12/F12 methods with medium-sized standard basis sets have been made more applicable after the introduction of a complementary auxiliary basis set (CABS). [85, 86] The CABS are used in addition to the orbital basis for insertion of the RI approximation.

The methodology and techniques discussed above have succesfully been used at the MP2 and CCSD level of theory and detailed reviews exist in the literature. [6, 78, 87] In this Thesis, an introductory description of the MP2-F12 method will be given. The reason is that, if conventional MP2 is the first step of the interference-corrected explicitly-correlated MP2 (which will be described in detail in the next chapters), the explicitly-correlated MP2 can be considered as the second step. In addition, the separability of the explicitly-correlated equations, intermediates and energy terms from those of the conventional methods make straightforward the description of the new method.

In the MP2-F12 theory the set of conventional double excitations ($a_a^\dagger a_i a_b^\dagger a_j$ from Eq. (2.34b)) is augmented by the excitations to the complementary virtual space. Both conventional $|ab\rangle$ and F12 pair basis form together the extended or complete virtual space (Table 2.1). The additional pair functions (or geminals) depend explicitly on r_{12} and they have the general form

$$|w_{xy}\rangle = \hat{Q}_{12}f_{12}\hat{S}_{xy}|\phi_x\phi_y\rangle, \tag{2.48}$$

where $|\phi_x\phi_y\rangle$ is a two-electron determinant and the rational generator $\hat{S}_{xy}$ ensures that the s- and p-wave coalescence conditions are satisfied [75]

$$\hat{S}_{xy}\varphi_x(1)\varphi_y(2)\sigma_x(1)\sigma_y(2) = \left(\frac{3}{8}\varphi_x(1)\varphi_y(2) + \frac{1}{8}\varphi_y(1)\varphi_x(2)\right)\sigma_x(1)\sigma_y(2). \tag{2.49}$$

In the last equation φ_x and σ_x are the spatial and spin components of a spin orbital ϕ_x, respectively (Table 2.1). In the extended virtual space α, β, double excitations into the geminals can be expressed as

$$|w_{xy}\rangle = \sum_{\alpha > \beta} w_{\alpha\beta}^{xy} a_\alpha^\dagger a_\beta^\dagger |vac\rangle,$$ (2.50)

with the overlap matrix elements $w_{\alpha\beta}^{xy} = \langle \alpha\beta | w_{xy}\rangle$ and the vacuum state $|vac\rangle$. The physical meaning of the geminals can be given as the linear combination of double excitations into spin orbitals of the complete basis. In Eq. (2.48), the strong-orthogonality projector $\hat{Q}_{12}$ can be defined differently, depending on the extend of the desirable virtual space. The most accurate description is known as ansatz 2 and is defined as

$$\hat{Q}_{12} = (1 - \hat{O}_1)(1 - \hat{O}_2)(1 - \hat{V}_1\hat{V}_2),$$ (2.51)

where $\hat{O}_i$ projects onto the space of (active and frozen) occupied and $\hat{V}_i$ onto the space of active virtual orbitals

$$\hat{O}_1 = \sum_i |\phi_i(1)\rangle\langle\phi_i(1)|, \qquad \hat{V}_1 = \sum_a |\phi_a(1)\rangle\langle\phi_a(1)|.$$ (2.52)

Thus, $\hat{Q}_{12}$ projects out of the functions $f_{12}|xy\rangle$ any contributions of occupied pairs $|ij\rangle$ or singly excited pairs $|i\alpha\rangle$, which ensures that the excitations into geminals are pure double excitations and have no Pauli-forbidden components. Furthermore, the factor $(1 - \hat{V}_1\hat{V}_2)$ keeps the geminals orthogonal to all conventional doubly excited pairs $|ab\rangle$. This minimizes the coupling between amplitudes for conventional and explicitly correlated double excitations and guarantees that approximations made in evaluating the geminal contributions will not change the basis set limit of a wave function method. This term also enforces the strong couplings between the two sets of double excitations and expressions for the geminal contributions to the wave function to vanish for ansatz 2.

In order to obtain the MP2-F12 energy, we have to define a new operator $\hat{T}_{2'}^{(1)}$ which constructs the double excitations in the complete virtual space. The first-order wave function (Eq. (2.35)) can be re-written as

$$|\Psi_{\text{F12}}^{(1)}\rangle = (\hat{T}_2^{(1)} + \hat{T}_{2'}^{(1)})|\text{HF}\rangle$$ (2.53)

with

$$\hat{T}_{2'}^{(1)} = \sum_{\substack{i>j \\ x>y}} c_{xy}^{ij} \sum_{\alpha>\beta} w_{\alpha\beta}^{xy} a_\alpha^\dagger a_i a_\beta^\dagger a_j.$$ (2.54)

The additional variable parameters in the F12 theory are the first-order coefficients c_{xy}^{ij}. The MP2-F12 correlation energy is then given by extending Eq. (2.36)

$$\delta E_{\text{MP2-F12}} = \langle \text{HF}|\hat{\Phi}|\Psi_{\text{F12}}^{(1)}\rangle = \langle \text{HF}|\hat{\Phi}(\hat{T}_2^{(1)} + \hat{T}_{2'}^{(1)})|\text{HF}\rangle$$
$$= \sum_{\substack{i>j \\ a>b}} t_{ab}^{ij} g_{ij}^{ab} + \sum_{\substack{i>j \\ x>y}} c_{xy}^{ij} V_{ij}^{xy}$$ (2.55)

The first term of the right-hand side of Eq. (2.55) corresponds to the conventional MP2 energy of Eq. (2.36). The second term consists of the additional F12 energy, which it is the product between the c_{xy}^{ij} coefficients and the V intermediate, whose matrix elements are of the form

$$V_{rs}^{xy} = \langle rs | r_{12}^{-1} | w_{xy}\rangle.$$ (2.56)

The first-order amplitudes t_{ab}^{ij} and the coefficients c_{xy}^{ij} are obtained after solving the equations

$$0 = \sum_c \left(t_{ac}^{ij} f_{cb} + t_{cb}^{ij} f_{ca} \right) - \sum_k \left(t_{ab}^{ik} f_{kj} + t_{ab}^{kj} f_{ki} \right) + \sum_{x>y} C_{ab}^{xy} c_{xy}^{ij} + g_{ab}^{ij} \tag{2.57}$$

$$0 = \sum_{v>w} B_{xy}^{vw} c_{vw}^{ij} - \sum_{v>w} X_{xy}^{vw} \sum_k \left(c_{vw}^{ik} f_{kj} + c_{vw}^{kj} f_{ki} \right) + \sum_{a>b} C_{xy}^{ab} t_{ab}^{ij} + V_{ij}^{xy} \, . \tag{2.58}$$

The computational cost scales as $\mathcal{O}(\mathcal{N}^6)$ with the system size, in contrast to the conventional MP2 calculation which scales as $\mathcal{O}(\mathcal{N}^5)$. The computational most demanding steps are the calculation of the B, V, X and C intermediates. These intermediates are of the form

$$V_{rs}^{xy} = \langle rs|r_{12}^{-1}|w_{xy}\rangle = \hat{S}_{xy}\langle rs|r_{12}^{-1}\hat{Q}_{12}f_{12}|xy\rangle \tag{2.59}$$

$$C_{ab}^{xy} = \langle ab|\hat{F}^{(0)}|w_{xy}\rangle = \hat{S}_{xy}\langle ab|(\hat{F}_1^{(0)} + \hat{F}_2^{(0)})\hat{Q}_{12}f_{12}|xy\rangle \tag{2.60}$$

$$B_{vw}^{xy} = \langle w_{vw}|\hat{F}^{(0)}|w_{xy}\rangle = \hat{S}_{vw}\hat{S}_{xy}\langle vw|f_{12}\hat{Q}_{12}(\hat{F}_1^{(0)} + \hat{F}_2^{(0)})\hat{Q}_{12}f_{12}|xy\rangle \tag{2.61}$$

$$X_{vw}^{xy} = \langle w_{vw}|w_{xy}\rangle = \hat{S}_{vw}\hat{S}_{xy}\langle vw|f_{12}\hat{Q}_{12}f_{12}|xy\rangle \, . \tag{2.62}$$

$\hat{F}^{(0)}$ is the zeroth-order contribution to the Fock operator

$$\hat{F}^{(0)} = \sum_{pq} f_{pq} a_p^\dagger a_q \, . \tag{2.63}$$

The most involved intermediate is B and together with V, they determine the leading explicitly correlated contributions to the wave function and energy.

An alternative approximate approach is to keep the c_{xy}^{ij} fixed to coefficients determined by the s- and p-wave coalescence conditions (Eq. (2.49)) at the interelectronic cusp at $c_{xy}^{ij} = \delta_{ix}\delta_{jy} - \delta_{iy}\delta_{jx}$. This approximation is called SP approach (or fixed-amplitudes) and it is less computational demanding because the optimization of the coefficients is skipped.

MP2-F12 enables to reach the basis set limit of the MP2 method by using medium-size basis sets. However, it is clear that by calculating the basis set limit of a method that cannot provide an accurate description of a system (total energies, interaction energies, derivatives or other properties), will lead to a wrong result. In order to use the full potential of the explicitly-correlated methods, the F12 technology should be used with highly accurate methods, such as coupled-cluster. For that reason, much recent activity has focused on the development of practical approximations of CCSD-F12.

It is beyond the scope of this chapter to give a detailed analysis of the equations and intermediates of this method. On the other hand, it is of high importance to highlight the applicability and accuracy obtained from computationally efficient formulations of the CCSD-F12 models. One of these models is the CCSD(R12) of Fliegl et al. [88,89] which was generalized to the F12 variant by Tew et al. [8] and implemented in the TURBOMOLE program package [90]. In particular, they had showed in Ref. [8] that CCSD(F12) with perturbative triples from a conventional CCSD(T) calculation [3] can reach accuracy close to the basis set limit of CCSD(T) (quintuple-zeta quality) by using a triple-zeta orbital basis set. Thus, this model (named as CCSD(T)(F12)) has been used in the present study for obtaining reference values.

CHAPTER 3

Interference-corrected MP2-F12

3.1 Hypothesis

CBS model chemistries have been successfully applied in various theoretical studies including among others thermochemistry, reaction barriers, and bond dissociation energies. The main idea of the present study is to merge the potential of the explicitly-correlated technology at the MP2 level with the effectiveness of the pair energy extrapolations, which constitutes the basis of the CBS model chemistries.

The methodology of the CBS models has been described in Section 2.2. The core of these methods is the pair energy extrapolation through the interference effect between the second- and infinite-order pair energies, as show in Eq. (2.21). The basis set limit of the second-order pair energies is obtained from two-point fits (Eqs. 2.20). These equations have been derived from the asymptotic behavior of the helium atom and they provide a reasonable accuracy for every electron pair, especially in strong interacting systems (*vide infra*).

The hypothesis made in this Thesis is that a more efficient, accurate and compact scheme can be formulated by obtaining the second-order pair energy basis set limit from the explicitly-correlated second-order perturbation theory. In other words, the second-order basis set limit is not any more a fitting parameter, but is calculated from MP2-F12. At the same time, application of the interference factor to the second-order truncation error leads to an estimation of the higher-order basis set limit. As a higher-order method, the CCSD(T) is used and results are calibrated with CCSD(T)(F12). It should be noted that this applies for correlation energies rather than total energies, as far as the HF limit shows a different convergence behavior.

Central to the CBS methods is Eq. (2.21). This can be modified based on the above as:

$$E_{\text{CCSD(T)(F12)}} \approx E_{\text{CCSD(T)}} + \sum_{i<j} F^{ij}[e_{ij}^{\text{MP2-F12}} - e_{ij}^{\text{MP2}}] \tag{3.1}$$

By keeping a consistent nomenclature with the previous chapter, ij pair energies are written with a small e_{ij}, while total energies with a capital E and correlation energies as δE. Note that the δe_{ij} term corresponds to pair energy truncation errors. For simplifying the notation, the interference factor $\left(\sum_{\mu=1}^{N} C_\mu\right)^2$ is written as F^{ij}. Summation over all pair energies is considered and thus, the basis set limit of the CCSD(T) correlation energy can be approximated as the sum of the CCSD(T) correlation energy obtained in a truncated basis and the interference-corrected second-order basis set truncation error

$$F^{ij}\delta e_{ij}^{(2)} = F^{ij}[e_{ij}^{\text{MP2-F12}} - e_{ij}^{\text{MP2}}]. \tag{3.2}$$

Interference-Corrected MP2-F12

obtain SCF canonical orbitals with **symmetry** or use **localization** schemes

in **subroutine** `rir12mp2`

do loop over memory partitions for *all* pairs   *in* **subroutine** `bvdirect`
    if (localization scheme) transform t_{ab}^{ij}
    write amplitudes in file   *subroutine* `rir12_intgett`
end do
write e_{ij}^{F12} pair energies in array   *subroutine* `rir12_intpcp`

subroutine `rir12_intcorr` (included in `rir12mp2`)

do loop over $\alpha\alpha$, $\beta\beta$, $\alpha\beta$-interorbital and $\alpha\beta$-intraorbital pairs
    do loop over memory partitions for *each* pair
       read amplitudes from file
       build $\mathbf{T}^{ij}$ matrix
    end do
    diagonalize $\mathbf{T}^{ij}$ with SVD
    calculate F^{ij} factors   (Eq. (3.13))
    calculate e_{ij}^{INT} pair energies   (Eq. (3.15))
end do
sum δE_{INT} and δE_{F12} energy terms

if (ccsd(t)) **then**
    calculate $E_{CCSD(T)}$
    add 2nd-order energy corrections to $E_{CCSD(T)}$ (Eq. (3.17))
end if

Algorithm 1: Pseudocode for the calculation of the second-order interference correction.

The right-hand side of Eq. (3.2) consists of a second-order correction to the CCSD(T) energy, which leads to the basis set limit of the coupled-cluster energy (Eq. (3.1)).

In the next sections, the development of the above method will be analyzed (Section 3.2) as it has been implemented in the RICC2 program of a working version of the TURBOMOLE program package [90]. The steps of the method are shown on the Algorithm 1. Illustrative examples that verify the applicability of the method will be discussed in Section 3.3. Finally, in Section 3.4 an alternative formulation based on different basis sets between the computational demanding CCSD(T) calculation (small basis) and the MP2-F12 part (large basis set) will be presented.

3.2 Implementation

3.2.1 Computation of the interference factor

The methodology of the original CBS model chemistry was followed closely for the calculation of the interference factor. The significance of the PNOs on the construction of the interference factor has been discussed in Section 2.2.1. In the PNO formalism each electron pair is treated by the most rapidly converging expansion of external orbitals like for example, the natural orbitals for the specific pair. [91] NOs are compressing the correlated wave function expansion and they are formed from the eigenvectors of a correlated density matrix. [92] The NOs are global to the system and they fail to include the short-range nature of the electron correlation. On the contrary, PNOs are taking into account this issue, as introduced by Edmiston and Krauss. [93] Therefore, as a first approach to the interference-corrected MP2-F12, the pair natural orbitals were used. These are constructed from the MP2 amplitudes (Eq. (2.39)) for every pair of active electrons ij. From these amplitudes, the pair density $\mathbf{D}^{ij}$ is calculated and with diagonalization the MP2-PNOs are obtained (the overbar denotes PNOs)

$$\mathbf{D}^{ij}\mathbf{d}_{\bar{a}}^{ij} = \bar{n}_{\bar{a}}^{ij}\mathbf{d}_{\bar{a}}^{ij}\,. \tag{3.3}$$

The $\bar{n}_{\bar{a}}^{ij}$ are the natural orbital occupation numbers. The MP2-PNO for a given ij pair is an expansion in terms of virtual MOs ($\langle a|$)

$$\langle a|^{ij} = \sum_{a} d_{a\bar{a}}^{ij}\langle a|\,. \tag{3.4}$$

There was recently an attempt for revival of the PNO methods in combination with localization schemes, either within the MP2-F12 framework [94], or with the coupled-electron pair approximation (CEPA) [95–97]. Another recent development on theories using a noncanonical virtual space constitutes the orbital-specific-virtual MP2 method [98,99] and its local CCSD variant [100]. It is based on the decomposition of the second-order amplitudes and is closely related to the PNO methods.

The interference factor is calculated from the squared sum of the amplitudes of the pair CI wave function

$$\Psi_{ij}^{\mathrm{CI}} = \sum_{ab}\mathbf{A}_{ab}(i,j)\Phi_{ij}^{ab}\,. \tag{3.5}$$

Φ_{ij}^{ab} is a configuration in which the electron pair ij is doubly excited to the virtual space (ab) and $\mathbf{A}$ is a variationally optimized coefficient. An expansion of the Φ_{ij}^{ab} as antisymmetrized products of one-electron orbitals allows us to focus on the two-electron term

$$\psi_{ij}^{\mathrm{CI}}(1,2) = \sum_{ab}\mathbf{A}_{ab}(i,j)\phi_a(1)\phi_b(2)\,, \tag{3.6}$$

or, in matrix form

$$\psi_{ij}^{\mathrm{CI}} = \phi\cdot\mathbf{A}\phi\,. \tag{3.7}$$

$\mathbf{A}$ is a square matrix which dimensions span the space of the virtual orbitals for the specific ij pair. By finding the proper unitary transformation, the coefficient matrix $\mathbf{A}$ is diagonalized. This unitary transformation is applied on the set of orbitals $\{i, j, \{a, b\}\}$ in order to get the PNOs. In other words, the CI wave function will determine the molecular PNOs and from the squared sum of the diagonal values of the coefficient matrix the interference factor will be obtained.

The construction of the PNOs for the calculation of the interference factor was skipped, as long as not all the terms of Eq. (3.3) are needed. In addition, the explicit calculation of the CI wave functions is a time-consuming procedure. For that reason, an alternative approach was used, based on the first-order wave function. Thus, the coefficient matrix $\mathbf{T}$ (instead of $\mathbf{A}$) is constructed from the t_{ab}^{ij} amplitudes of the second-order Møller-Plesset perturbation theory. The second-order amplitudes still hold all the information for the double electron excitations to the virtual space. They represent the probability for the excitation of a pair of electrons from occupied orbitals i, j to virtual orbitals a, b. Like before, the matrix $\mathbf{T}$ is a matrix over pairs of virtual orbitals ab and is different for each pair of i, j.

The procedure for obtaining the interference factors varies for the triplet ($\alpha\alpha$ and $\beta\beta$) pairs and for the $\alpha\beta$-intraorbital and $\alpha\beta$-interorbital pairs. Therefore, they will examined separately:

1. Triplet pairs

   For every $\sigma\sigma$ pair, where σ is either α or β, the orbitals of the $i\sigma j\sigma$ electrons have an antisymmetric spatial component. This means that $t_{ab}^{ij} = -t_{ba}^{ij}$ and the $\mathbf{T}^{i\sigma j\sigma}$ amplitudes matrix is antisymmetric. For a given virtual ab pair, the matrix element of $\mathbf{T}^{i\sigma j\sigma}$ is of the form

$$(T^{i\sigma j\sigma})_{ab} = \frac{1}{2} \frac{(ia|jb) - (ib|ja)}{\varepsilon_i + \varepsilon_j - \varepsilon_a - \varepsilon_b}. \tag{3.8}$$

   The diagonal elements are zero, since double $\alpha\alpha$ or $\beta\beta$ excitations to the same orbital are forbidden (Pauli exclusion principle). Thus, the total number of matrix elements of the $\mathbf{T}^{i\sigma j\sigma}$ matrix can be reduced to $\binom{V}{2}$ unique entries, where V is the number of virtual orbitals. An ordinary symmetric matrix has the same number of unique elements and requires only one unitary transformation for the reduction to V coefficients (diagonal elements)

$$\mathbf{d} = \mathbf{U}^{\mathrm{T}} \, \mathbf{T}^{i\sigma j\sigma} \, \mathbf{U}. \tag{3.9}$$

   Since the diagonal elements are zero, the reduced second-order amplitudes matrix $\mathbf{d}$ cannot be diagonal. In order to get the $\mathbf{d}$ matrix to its diagonal form, the one-electron density matrix is diagonalized as

$$\mathbf{D}^{i\sigma j\sigma} = (\mathbf{T}^{i\sigma j\sigma} \, (\mathbf{T}^{i\sigma j\sigma})^{\mathrm{T}})^{\frac{1}{2}}. \tag{3.10}$$

   The iterative Jacobi method was tested for the determination of the diagonal elements. Thus, the interference factor is obtained from the trace of the diagonal semi-definite $\mathbf{D}^{i\sigma j\sigma}$ matrix as

$$F^{i\sigma j\sigma} = \{1 - \mathrm{tr}(\mathbf{D}^{i\sigma j\sigma})\}^2 \tag{3.11}$$

   A different strategy to obtain the diagonal elements is by applying the singular-value-decomposition (SVD) method

$$\mathbf{T}^{i\sigma j\sigma} = \mathbf{V} \, \Lambda^{i\sigma j\sigma} \, \mathbf{W}^{\mathrm{T}}. \tag{3.12}$$

   In the SVD method, the $\mathbf{T}^{i\sigma j\sigma}$ is written as the product of an orthogonal matrix $\mathbf{V}$, a square diagonal matrix $\Lambda^{i\sigma j\sigma}$ with positive or zero elements and the transpose of the square matrix $\mathbf{W}$. The singular values of $\mathbf{T}^{i\sigma j\sigma}$, i.e. the diagonal elements of $\Lambda^{i\sigma j\sigma}$, are used for the calculation of the interference factor. For the SVD method, the DGEBRD and DBDSQR subroutines of the LAPACK software package were used. [101] In line with Eq. (3.11), the interference factor is computed from

$$F^{i\sigma j\sigma} = \{1 - \mathrm{tr}(\Lambda^{i\sigma j\sigma})\}^2 \tag{3.13}$$

   Both Jacobi and SVD methods gave the same values for the interference factors. However, the SVD is superior to the Jacobi method, not only because it is much faster, but also due to the fact that it can handle non-square matrices. This proved to be important for the unrestricted code and for the singlet pairs. In this case, the $\mathbf{T}$ matrix has $v_\alpha \times v_\beta$ dimensions, where v_α is the number of virtual orbitals with α spin and v_β is the number of virtual orbitals with β spin.

2. Singlet intraorbital pairs
   In the singlet intraorbital case, the α and β electrons are spatially equivalent, so that excitations of the form $ii \rightarrow ab$ are physically the same as $ii \rightarrow ba$. The corresponding $\mathbf{T}^{i\alpha j\beta}$ amplitudes matrix is a symmetric matrix with elements

$$(T^{i\alpha j\beta})_{ab} = \frac{(ia|jb)}{\varepsilon_i + \varepsilon_j - \varepsilon_a - \varepsilon_b}. \tag{3.14}$$

The number of unique elements in this case is $\binom{V}{2} + V$, where V is the number of virtual orbitals. The additional V elements correspond to the diagonal non-zero amplitudes. The interference factor $F^{i\alpha j\beta}$ is computed again from the singular values of the SVD method (Eqs. (3.12) and (3.13)).

3. Singlet interorbital pairs
   The excitation of an $\alpha\beta$ pair in the virtual space has an equal mixture of singlet and triplet components. This has as a result that the t_{ab}^{ij} amplitudes matrix $\mathbf{T}$ is nonsymmetric and the excitations $ij \rightarrow ab$ and $ij \rightarrow ba$ are not physically equivalent. The total number of unique entries is V^2. Again, the SVD was used.

The above issues can be extended for unrestricted cases straightforwardly from the SVD method. The only parameters that should be taken into account are the dimensions of the involved matrices, as far as the number of α virtual orbitals is not the same with those of β virtual orbitals ($v_\alpha \neq v_\beta$). This holds especially for the singlet pairs, where the dimensions of the $\mathbf{T}^{i\alpha j\beta}$ matrix are $v_\alpha \times v_\beta$. Therefore, for the SVD method, the matrix $\mathbf{V}$ is a column-orthogonal matrix ($v_\alpha \times v_\beta$), while $\Lambda^{i\sigma j\sigma}$ and $\mathbf{W}$ are $v_\beta \times v_\beta$ square matrices.

Once the interference factors F^{ij} are known, the second-order interference energy correction (INT) is computed for each pair

$$e_{ij}^{\text{INT}} = (F^{ij} - 1) \, [e_{ij}^{\text{MP2-F12}} - e_{ij}^{\text{MP2}}] = (F^{ij} - 1) \, e_{ij}^{\text{F12}} \tag{3.15}$$

since the MP2-F12 correlation energy can be decomposed into the conventional and the F12 components (Eq. (2.55)). The e_{ij}^{F12} term can be computed either with the fixed or invariant approach. In the current status of the code, in the output are printed pair energies from both approximations A and B. These will be mentioned as $e_{ij}^{\text{F12/A}}$ and $e_{ij}^{\text{F12/B}}$, with the corresponding INT terms $e_{ij}^{\text{INT/A}}$ and $e_{ij}^{\text{INT/B}}$, respectively. The INT pair energies are summed up as follows

$$\delta E_{\text{INT/A}} = \sum_{i<j} (F^{ij} - 1) \, e_{ij}^{\text{F12/A}}, \qquad \delta E_{\text{INT/B}} = \sum_{i<j} (F^{ij} - 1) \, e_{ij}^{\text{F12/B}}. \tag{3.16}$$

The (positive) energies $\delta E_{\text{INT/A}}$ and $\delta E_{\text{INT/B}}$ correspond to what is denoted as $\Delta E(\text{CBS-int})$ in the CBS model chemistries.

Two different options are offered at the current status of the code. The first computes only the MP2-F12 energy with the INT correction and it is initialized from the mp2 energy only flag. The second includes by the same ricc2 call the CCSD(T) calculation (ccsd(t) flag). It starts right after the end of the interference-corrected explicitly-correlated MP2 calculation (INT-MP2-F12), after deleting first the no longer important files with the F12 intermediates. At the end, as final output the approximate CCSD(T)/CBS energy is printed

$$E_{\text{CCSD(T)/CBS}} \approx E_{\text{CCSD(T)}} + \delta E_{\text{INT}} + \delta E_{\text{F12}} \,. \tag{3.17}$$

Both flags should be written in the `control` file under the `$ricc2` keyword and should be followed from the `intcorr` flag which will initialize the INT correction. A detailed output is also available (`intcorr all` flag) which prints a list with all the calculated interference factors and the corresponding $e_{ij}^{\text{INT/A}}$ and $e_{ij}^{\text{INT/B}}$ pair energies.

From now on, this method will be abbreviated as "CCSD(T)-INT-F12". This name was chosen because it includes in it all the three energy components of Eq. (3.17) which constitutes the estimate to the CCSD(T) basis set limit.

3.2.2 Technical details

In this subsection, some technical considerations for the optimum performance of the interference-corrected MP2-F12 method will be discussed and techniques which allow its extension to larger systems will be presented. These have to do mainly with two issues.

The first is about the memory allocation of the corresponding arrays which allowed the construction of an efficient code, being able to handle large molecular complexes with many basis set functions. For example, a calculation of a T-shaped pyrrole-benzene complex with the large YP-aug-cc-pV5Z basis set (2277 basis functions) was made feasible only after splitting the virtual space into smaller memory partitions.

The second consideration has to do with the choice of the starting SCF orbitals, for either restricted or unrestricted cases. It was already known from the original work of the asymptotic extrapolations of the CBS model chemistries that the interference effect is orbital variant. The reason is that, even if the sum of the second-order pair energies is orbital invariant (i.e. the $e_{ij}^{\text{MP2}}(N)$ and $e_{ij}^{\text{MP2}}(\infty)$ terms in the CBS equations or the e_{ij}^{MP2} and $e_{ij}^{\text{MP2-F12}}$ terms in the current formulation, respectively) the interference factor is sensitive to the choice of the initial (active occupied and virtual) orbitals. For that reason, in order to make the method uniquely defined, a specific protocol should be followed. This has been achieved with orbital alignment, symmetry restrictions (when possible) and localization schemes.

a. Memory Partitions

The design of the interference-corrected MP2-F12 method has been done in line with the RICC2 program of the TURBOMOLE program package. The general philosophy of the RICC2 program is the application to large systems and thus, efficient procedures are adopted and routines are designed with low disk and memory requirements. Density fitting is used for all two-electron integral evaluations and four-index integrals are evaluated at the point they are required and are not stored on disk. [102] The new method presented here is in line with these considerations.

The main subroutine that controls all the steps of the conventional and explicitly-correlated MP2 is `rir12mp2`. After the calculation of all two- and three-index quantities, the semi-direct algorithm of subroutine `bvdirect` calculates the most important intermediates of MP2-F12 directly from three-index quantities, like the **B**, **V** and similar matrices, while it avoids the time-consuming

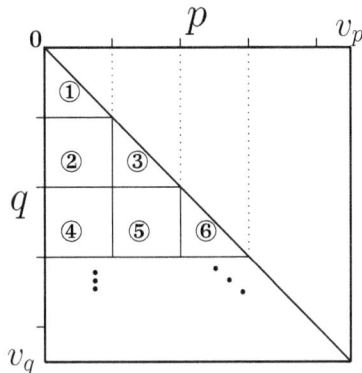

Figure 3.1: Memory partitions of the pq virtual space.

reading and writing of four-index integrals. In these loops over different intermediates and contraction schemes, the second-order amplitudes are also calculated by antisymmetrized singlet and triplet quantities and stored on disk from the newly introduced subroutine `rir12_intgett` (see Algorithm 1). The disk storage requirements are O^2V^2, where O is the number of active occupied orbitals and V the number of virtual orbitals.

In order to reduce the amount of data kept in memory during the aforementioned loops, the pq virtual space is partitioned into smaller parts. These memory partitions are organized as follows: the spatial orbital indices run over the triangle $p \leq q$ and the square i, j. The indices i, j are antisymmetrized to yield the t_{ab}^{ij} amplitudes (in `rir12_intcorr`). Then, either only the q dimension or both pq dimensions split in batches in order to build smaller arrays which can be kept in memory. These partitions follow the triangular $p \leq q$ form and they can have either triangular or rectangular shape, as it is shown on Figure 3.1. On the triplet quantities, $p < q$ and the diagonal elements are not included on the partitions (diagonal elements are zero). On the contrary, the diagonal elements have non-zero values for the singlet quantities.

The number of batches into the q and/or p dimensions will split depends on some parameters. The most important parameter is the maximum memory set by the user (keyword `$maxcor` of the `control` file). If maximum memory is less than the one needed for the allocation of all the arrays, the algorithm will split the virtual space into partitions. A second important criterion is the number of processors. This serves for an efficient distribution over the available resources for a parallel calculation. Other parameters which affect the partitioning are the total number of basis functions, the number of ij pairs and the number of intermediates needed by the F12 code. The latter has to do with the approximations which the user wants to apply to the explicitly-correlated part of MP2.

After the end of the loops over all memory partitions, the computation of the interference factor begins for every electron pair (subroutine `rir12_intcorr`). For this, a nested loop is set up, with an outer loop over all different electron pairs, and an inner loop over the partitions. The singlet and triplet antisymmetrized elements are read and the $\mathbf{T}^{ij}$ matrix is built (Algorithm 1). The procedure described in Subsection 3.2.1 is followed, depending on the spin of the ij electron pairs.

An important note is that the algorithm followed for the calculation of the INT energy correction is almost free of cost. The t_{ab}^{ij} second-order amplitudes are already computed for the conventional MP2, which is part of the MP2-F12 code. The only time spend is on the I/O of the files where the antisymmetrized vectors are stored.

$$
\begin{array}{c}
\begin{array}{cccc}
2s & 2p_z & 2p_x & 2p_y
\end{array} \\
\begin{array}{c}
2s \\
2p_{x+y} \\
2p_{x-y}
\end{array}
\left(
\begin{array}{cccc}
1.00 & 0.0 & 0.0 & 0.0 \\
0.0 & 0.0 & -0.76 & 0.65 \\
0.0 & 0.0 & 0.65 & 0.76
\end{array}
\right)
\end{array}
\Rightarrow
\begin{array}{c}
\begin{array}{cccc}
2s & 2p_{x+y} & 2p_{x-y} & 2p_z
\end{array} \\
\begin{array}{c}
2s \\
2p_{x+y} \\
2p_{x-y}
\end{array}
\left(
\begin{array}{cccc}
1.00 & 0.0 & 0.0 & 0.0 \\
0.0 & 1.00 & 0.0 & 0.0 \\
0.0 & 0.0 & 1.00 & 0.0
\end{array}
\right)
\end{array}
$$

Figure 3.2: Orbital alignment of the $2p$ orbitals of the fluorine atom.

b. Orbital Alignment

In open-shell molecules with high symmetry, the α spin orbitals are not uniquely related to the β spin orbitals. For example, the three occupied α spin $2p$ orbitals of the ground state of the fluorine atom have collectively spherical symmetry, while the two β spin $2p$ orbitals must point in arbitrary directions. Although the MP2 and the F12 energies are invariant to unitary transformations, the INT energy will vary. This was already noted in the original CBS model chemistry papers [43] and as a possible solution, the authors had proposed to fix the relative orientation of α and β spin orbitals.

This suggestion was also tested in the present study, in order to formulate a *uniquely defined open-shell interference correction*. A straightforward procedure is obtained from the diagonalization of the $\alpha\beta$-overlap matrix

$$
^{\alpha\beta}\mathbf{S}_{\mathrm{MO}} = {}^{\alpha}\mathbf{C}_{\mathrm{MO}}^{\dagger}\,\mathbf{S}_{\mathrm{AO}}\,{}^{\beta}\mathbf{C}_{\mathrm{MO}} \tag{3.18}
$$

where $\mathbf{S}_{\mathrm{AO}}$ is the overlap matrix for the AO basis set and $^{\alpha}\mathbf{C}_{\mathrm{MO}}$ and $^{\beta}\mathbf{C}_{\mathrm{MO}}$ the α- and β-spin occupied molecular orbital coefficients. The diagonalization of the $^{\alpha\beta}\mathbf{S}_{\mathrm{MO}}$ rectangular matrix has been done with the SVD method, employing two different unitary transformations, $^{\alpha}\mathbf{U}$ and $^{\beta}\mathbf{U}$. The application of these unitary transformations to the molecular coefficients will result in a new set of transformed orbitals $^{\alpha}\mathbf{C}'_{\mathrm{MO}}$ and $^{\beta}\mathbf{C}'_{\mathrm{MO}}$

$$
\begin{aligned}
^{\alpha}\mathbf{C}'_{\mathrm{MO}} &= {}^{\alpha}\mathbf{C}_{\mathrm{MO}}\,{}^{\alpha}\mathbf{U} \\
^{\beta}\mathbf{C}'_{\mathrm{MO}} &= {}^{\beta}\mathbf{C}_{\mathrm{MO}}\,{}^{\beta}\mathbf{U}
\end{aligned}
\tag{3.19}
$$

and thus the transformed α and β spin molecular orbitals are *aligned*.

An illustrative example of the application of the orbital alignment is the aforementioned case of the fluorine atom. Before the orbital transformation, the $2p_x$ and $2p_y$ have different directions for the α and β spins (Figure 3.2, left side). After the application of the unitary transformations of Eqs. (3.19) the occupied β orbitals will be orthogonal on the α spin $2p_z$ orbital but aligned with the other two (Figure 3.2, right side). In energy terms, this orbital alignment slightly varies the INT energy correction. For example, for the fluorine atom, and with the cc-pVDZ-F12 basis set [103], the $E_{\mathrm{INT}/\mathrm{A}}$ without the orbital alignment is 12.59 mE_{h}, while with the application of the aforementioned transformations, it changes to 13.11 mE_{h}. For the CH$_3$ open-shell molecule, which has two degenerate orbitals due to its D_{3h} symmetry, the difference between with and without the orbital alignment is 0.07 mE_{h}.

c. SCF canonical orbitals with symmetry

A different approach to the orbital rearrangement of the open-shell cases is the application of symmetry restrictions to our starting orbitals. For example, for the fluorine atom, if SCF canonical orbitals are obtained under D_{2h} symmetry, the highest Abelian point group, then β orbitals are not any more in arbitrary directions. In energy terms, and for the interference-correction, this approach yields the same result with the orbital alignment of canonical orbitals obtained without symmetry restrictions.

Starting the post-HF calculations with symmetrical orbitals solves one extra problem which appears in the INT-MP2-F12 method. MP2 or F12 energy components are not affected by the choice of the starting orbitals. On the contrary, the calculation of the individual interference factors for each of those pairs heavily depends on the initial choice. The reason is that the t_{ab}^{ij} amplitudes are calculated from the i, j, a and b orbital energies and the corresponding g_{ij}^{ab} 2-electron integral and both are affecting the value of the interference factor. If symmetry restrictions are imposed, pair energies between electrons occupying degenerate orbitals are equivalent. This option offers a *unique* description of the e_{ij}^{INT} which maximizes the interference correction, which is by definition positive. Therefore, symmetry restrictions impose isoenergetic INT pair terms and lead to a faster convergence. All calculations presented in this work are started with symmetrical SCF orbitals, unless noted otherwise. Numerical results concerning this issue are given in Section 3.3.

d. Localization Schemes

In the original work on the CBS model chemistries, the issue of the starting orbitals was solved with the implementation of localization schemes. Apart from the benefits that the localized orbitals offer to a quantum chemistry calculation, like the description of MOs closer to chemical intuition or the elimination of the long-range interaction part of the virtual space [104], they offer a unique description to the interference correction (or second-order CBS extrapolation). The two available localization schemes of the RICC2 code (Boys [105, 106] and Pipek-Mezey [107]) were also used for closed-shell molecules in this study. The Boys procedure minimizes the spatial extension of the MOs by maximizing the squared distance of charged centers of different localized molecular orbitals (LMOs) while Pipek-Mezey is based on the localization of the Mulliken atomic charge distribution by maximizing Mulliken's gross orbital populations. If a localization scheme is chosen, then one extra step is added in the calculation of the interference factor; for every memory partition, the first-order amplitudes are transformed from the semi-canonical to the localized before they are stored on disk (Algorithm 1).

The Pipek-Mezey localization scheme is used successfully in the protocol of the CBS model chemistries. Although the applicability of the CBS methods for atomization energies, dissociation energies, ionization potentials, electron affinities or reaction barriers has been demonstarted in many applications, their effectiveness on weak interacting systems is still on debate. [7] A possible reason for this is the choice of the localization scheme which does not allow for a proper description of the MOs and thus an efficient convergence of the ij pair energies. For example, Boys localization for the neon atom will give four energetically degenerate sp-like valence LMOs, while Pipek-Mezey LMOs will be identical with the canonical MOs; one s and three degenerate s. This analysis causes no harm for the total energy of the atom (see also Chapter 3.3) but it leads to mistakes when two atoms are in a non-bonding condition and localization is used for the description of their MOs, like, for example, in the neon dimer. In fact, Boys localization yields the correct potential curve for this specific system. Further discussion about this issue will be given on Chapter 5.

Azar and Head-Gordon [108] recently proposed an energy decomposition analysis for weak interactions with LMOs at the CCSD level. In their method, they used the Boys localization for both occupied and virtual orbitals. Additionally, both Neese and co-workers [95] in their localized PNO methods and Tew and co-workers [94] in their local MP2-F12 theory with PNOs are using the Boys localization scheme. These applications of the Boys scheme can be considered as proofs of its better performance over the Pipek-Mezey scheme for methods based on PNOs.

3.3 Verification of the Method

3.3.1 CI convergence of the helium atom

The applicability of the second-order corrections from the interference-corrected MP2-F12 on the basis set convergence of the CCSD(T) energy was checked initially with some simple test case molecules. A formidable example is the ground state of the helium atom because it includes one and only one $1s^2 \alpha\beta$ electron pair. The CCSD(T)-INT-F12 results can then be compared either with the FCI energy or with extrapolated values to the basis set limit which have been constructed based on the helium atom, like the natural orbital (or principal) expansion which was described in Subsection 2.1.2.

Kong *et al.* [78] demonstrated the slow convergence of the principal expansion for the helium atom and the need of extrapolation schemes or explicitly-correlated wave functions (Table 3.1). A NO expansion based on Eq. (2.7) was used and the results were compared with the exact FCI value (-2.903724377 E_h). The basis set error (ΔE_N) is similar with the corresponding errors from the CI energies (ΔE_{CI}) computed with Dunning's family of basis sets cc-pVXZ [19]. However, in both cases, the main problem still remains the slow asymptotic rate of convergence. Even for the large sextuple-zeta basis set, the basis set truncation error still remains significantly large (~ 0.3 mE_h). The CCSD total energy (i.e. $E_{HF} + \delta E_{CCSD}$) also shows the same slow convergent behavior. (Note that for a 2-electron system there are no triple excitations, the energy from the perturbative triples is zero and thus, $E_{CCSD(T)}$ can be written also as E_{CCSD}, which is equivalent with the CI. The numerical differences on the Tables 3.1 and 3.2 between the CI and CCSD energies are due to the application of the RI on the CCSD calculations.)

Addition of the second order corrections from MP2-F12 (Eq. (3.17)) to the $E_{CCSD(T)}$ energy change the convergence behavior as the size of the basis is increased. All F12 and INT results showed on Tables 3.1 and 3.2 have been obtained from the invariant MP2-F12 within the approximation A and do not include the HF correction from CABS Singles [109]. Adding only the F12 component ($\Delta E_{CCSD+F12}$), the total energy is overestimated, leading to a similar slow convergence. However, there are two differences; firstly, increase of the size of the basis set increases the correlation energy and the FCI energy is approached from below and secondly, double- and triple-zeta energies are not consistent with the energies from the larger basis sets. This has to do mainly with the convergence of the SCF energy: by examining only the correlation energy (Table 3.2), the inverted asymptotic convergence is regained.

The INT component does not only correct the $\Delta E_{CCSD+F12}$ energy, but leads to a faster convergence of the higher-order method (CCSD or CI). The $\Delta E_{CI/cc-pV6Z}$ energy is already obtained from the CCSD-INT-F12 calculation with a quadruple-zeta quality basis set, which is in size the one third of the sextuple-zeta: 30 basis functions instead of 91. Indeed, for the correlation energy, perfect agreement with the FCI at the CBS limit is achieved with the cc-pV5Z. For this case also, the correlation energy obtained from the quadruple-zeta is in better agreement than the $\Delta E^{corr}_{CI/cc-pV6Z}$.

Table 3.1: Basis set convergence of the HF, CI, CCSD and second-order corrected CCSD energies (in E_h) for the ground state helium atom. Each cc-pVXZ correlation consistent basis set includes n_{bf} basis functions with angular momentum up to L_{max} and corresponds to the NO expansion set with $N = L_{max} + 1$.

Basis set	L_{max}	n_{bf}	E_{HF}	E_{CI}[a]	ΔE_{CI}[a]	ΔE_N[a,b]	ΔE_{CCSD}	$\Delta E_{CCSD+F12}$	$\Delta E_{CCSD-INT-F12}$
cc-pVDZ	1	5	-2.855160	-2.887595	0.016129	0.010977	0.016132	0.005182	0.007827
cc-pVTZ	2	14	-2.861153	-2.900232	0.003492	0.003115	0.003494	-0.000478	0.000901
cc-pVQZ	3	30	-2.861514	-2.902411	0.001313	0.001271	0.001316	-0.000511	0.000232
cc-pV5Z	4	55	-2.861625	-2.903152	0.000572	0.000645	0.000573	-0.000367	0.000051
cc-pV6Z	5	91	-2.861673	-2.903432	0.000292	0.000361	0.000292	-0.000261	0
			$\vdots$	$\vdots$	$\vdots$	$\vdots$			
complete	∞	∞	-2.86168^c	-2.903724	0	0			

[a] Reference [78]
[b] From Eq. (2.7)
[c] Reference [110]

Table 3.2: Basis set convergence of the correlation energy of CI, CCSD and second-order corrected CCSD energies (in E_h) for the ground state helium atom.

Basis set	L_{max}	n_{bf}	E_{CI}^{corr}[a]	ΔE_{CI}^{corr}[a]	ΔE_{CCSD}^{corr}	$\Delta E_{CCSD+F12}^{corr}$	$\Delta E_{CCSD-INT-F12}^{corr}$
cc-pVDZ	1	5	-0.032435	0.009609	0.009613	-0.001337	0.001307
cc-pVTZ	2	14	-0.039079	0.002965	0.002968	-0.001005	0.000374
cc-pVQZ	3	30	-0.040897	0.001147	0.001150	-0.000677	0.000066
cc-pV5Z	4	55	-0.041527	0.000517	0.000518	-0.000422	-0.000004
cc-pV6Z	5	91	-0.041759	0.000285	0.000285	-0.000268	-0.000007
			$\vdots$	$\vdots$			
complete	∞	∞	-0.042044	0			

[a] Reference [78]

3.3.2 The CBS limit of the neon atom

Barnes et al. [111] used as large as cc-pV10Z basis sets in order to establish the fc-CCSD(T)/CBS limit of the neon atom. They partitioned the correlation energy into two components for MP2 energy (summations over all ij pairs for opposite-spin $^{\alpha\beta}e_{ij}^{(2)}$ and same-spin $^{\alpha\alpha}e_{ij}^{(2)}$), a higher-order correction $E_{CCSD} - E_{MP2}$ and $E_{(T)}$ and they extrapolated separately each term. The CCSD limit was reached through calibration with CCSD-R12 [85, 112] and finite element MP2 results [113] while the extrapolated value of the perturbative triple excitations was obtained from asymptotic convergence of the form $(\ell + a)^{-3}$, with $a = -1.640$. Their CCSD(T)/CBS limit was $-128.869\,236 \pm 0.000\,02\,E_h$ and the corresponding correlation energy $-322.138\,\mathrm{m}E_h$.

The accuracy obtained from the conventional fc-CCSD(T) method with the interference-corrected MP2-F12 energy components is less than one milli-hartree from the CCSD(T)/CBS limit. cc-pVQZ-F12 basis set with canonical orbitals or Pipek-Mezey LMOs gave for the correlation energy a value of $-321.219\,\mathrm{m}E_h$, only $0.92\,\mathrm{m}E_h$ higher than the CBS limit. A Boys localization scheme yielded a correlation energy $0.07\,\mathrm{m}E_h$ higher than canonical or Pipek-Mezey LMOs.

For comparison, explicitly-correlated CCSD(T) with the same basis set gives $-321.299\,\mathrm{m}E_h$, still less than one milli-hartree difference from the CBS limit. Indeed, this is an indication that CCSD(T)(F12) can be used as a reference for controlling the accuracy of the composite method presented here. For demonstrating the significance of the above results, the non-extrapolated value obtained from the additivity of MP2-$\alpha\beta$ pair energies should be discussed. [111] The MP2-$\alpha\alpha$ pair energies, CCSD–MP2, and (T) components with the cc-pV10Z basis set ($-320.844\,\mathrm{m}E_h$) has a deviation of $1.29\,\mathrm{m}E_h$, which is larger than both CCSD(T) with the second-order corrections or the explicitly-correlated CCSD(T) results.

3.3.3 The interference factors for the N and CH$_4$ cases

But what exactly is gained through the interference effects in the pair energies and what is the role of the interference factors? These questions can be answered by looking at the F12/A and F12/B pair energies computed at the MP2-F12/A and MP2-F12/B levels, respectively, as well as the corresponding interference corrections for the nitrogen atom and the methane molecule (Tables 3.3 and 3.4). These systems were chosen for their high symmetry, and thus, their small number of distinct pair energies (degeneracy factors are included in the reported energies). Results are shown for two basis sets (cc-pVDZ-F12 and cc-pVTZ-F12). The interference factors are slightly smaller in the large basis set. At the same time, the $e_{ij}^{F12/A}$ and $e_{ij}^{F12/B}$ are also smaller: by increasing the basis set, the CBS limit is approached faster and thus, the second-order corrections should have a smaller contribution. It should be mentioned again that the essence of the (positive) e_{ij}^{INT} terms is to increase the second-order contributions from the second-order F12 component to the CCSD(T) energy. The plain E_{F12} of MP2-F12, when is added on a higher-order energy like $E_{CCSD(T)}$, tends to overestimate the correlation energy.

From Tables 3.3 and 3.4 is also clear that the F^{ij} for same-spin are significantly larger than for opposite-spin pairs. The above two observations highlight the need for separate extrapolations for each pair. However, in a recent study [114] a global scaling factor $f_{int} = 0.78$ was used, which was obtained by fitting calculated atomization energies to experimental data. Results obtained from that approach were in a good agreement with CCSD(T)(F12) energies, even if an individual scaling for the same- and opposite-spin pairs would have been superior. This suggestion is similar to the spin-component-scaled MP2 method (SCS-MP2) [115], but now only for the basis set truncation error.

Table 3.3: Interference factors and valence-shell MP2-F12 pair contributions (in mE_h) of the nitrogen atom in the cc-pVDZ-F12 and cc-pVTZ-F12 basis sets.

Pair	Spin	$e_{ij}^{F12/A}$	$e_{ij}^{F12/B}$	F^{ij}	$e_{ij}^{INT/A}$	$e_{ij}^{INT/B}$
cc-pVDZ-F12						
$2s2p$	$\alpha\alpha$	-1.684	-1.443	0.93056	0.117	0.100
$2p2p'$	$\alpha\alpha$	-3.022	-2.658	0.84157	0.479	0.421
$2s^2$	$\alpha\beta$	-2.934	-2.635	0.68680	0.919	0.825
$2s2p$	$\alpha\beta$	-15.639	-13.676	0.67332	5.109	4.468
Sum		**-23.278**	**-20.413**		**6.623**	**5.814**
cc-pVTZ-F12						
$2s2p$	$\alpha\alpha$	-0.520	-0.471	0.90682	0.048	0.044
$2p2p'$	$\alpha\alpha$	-0.754	-0.692	0.80899	0.144	0.132
$2s^2$	$\alpha\beta$	-1.541	-1.429	0.61838	0.588	0.545
$2s2p$	$\alpha\beta$	-6.378	-5.861	0.58390	2.654	2.439
Sum		**-9.193**	**-8.453**		**3.435**	**3.160**

Energies in mE_h, fc approximation.

Table 3.4: Interference factors and valence-shell MP2-F12 pair contributions (in mE_h) of the CH_4 molecule in the cc-pVDZ-F12 and cc-pVTZ-F12 basis sets.

Pair	Spin	$e_{ij}^{F12/A}$	$e_{ij}^{F12/B}$	F^{ij}	$e_{ij}^{INT/A}$	$e_{ij}^{INT/B}$
cc-pVDZ-F12						
$2a_1 1t_2$	$\sigma\sigma$	-1.188	-1.030	0.90173	0.117	0.101
$1t_2 1t_2'$	$\sigma\sigma$	-2.293	-1.996	0.82480	0.402	0.350
$2a_1^2$	$\alpha\beta$	-2.027	-1.748	0.62313	0.764	0.659
$2a_1 1t_2$	$\alpha\beta$	-14.408	-12.564	0.66177	4.873	4.250
$1t_2 1t_2'$	$\alpha\beta$	-10.333	-9.085	0.61040	4.026	3.540
$1t_2^2$	$\alpha\beta$	-8.822	-7.775	0.51084	4.316	3.804
Sum		**-39.071**	**-34.198**		**14.497**	**12.702**
cc-pVTZ-F12						
$2a_1 1t_2$	$\sigma\sigma$	-0.336	-0.305	0.88681	0.038	0.035
$1t_2 1t_2'$	$\sigma\sigma$	-0.536	-0.489	0.80122	0.107	0.097
$2a_1^2$	$\alpha\beta$	-1.065	-0.961	0.56160	0.467	0.421
$2a_1 1t_2$	$\alpha\beta$	-6.825	-6.199	0.59342	2.775	2.521
$1t_2 1t_2'$	$\alpha\beta$	-5.399	-4.910	0.55178	2.420	2.201
$1t_2^2$	$\alpha\beta$	-4.424	-4.050	0.44537	2.454	2.247
Sum		**-18.584**	**-16.914**		**8.260**	**7.521**

Energies in mE_h, fc approximation.

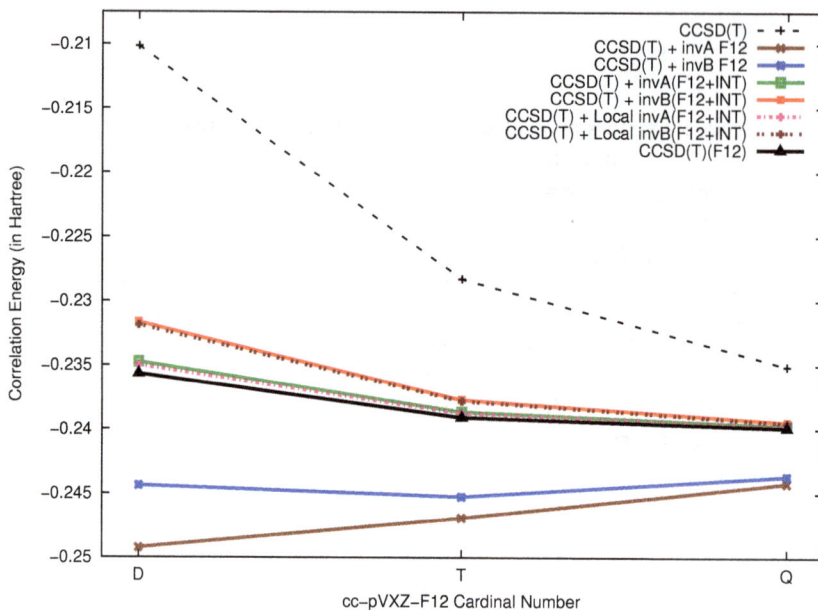

Figure 3.3: Conventional, corrected thought interference-corrected MP2-F12 and explicitly-correlated CCSD(T) correlation energy basis set convergence of the methane molecule.

An important issue which was under consideration in the present study is the lack of orbital invariance of the interference-corrected energy component, which has already been mentioned in Subsection 3.2.2. This discrepancy can be displayed with a high-symmetry molecule and thus, the methane molecule in the ground state was chosen. The choice of cc-pVDZ-F12 basis set [103] without any symmetry restrictions yields a δE_{INT} of 13.80 mE_h (for invariant MP2-F12 with approximation A). On the contrary, starting SCF orbitals with T_d symmetry results in a larger INT energy term (14.50 mE_h). Localization of the orbitals proved that the latter value is the one that should be accepted: the corresponding value of 14.23 mE_h is closer to the value obtained from SCF orbitals with symmetry than the one without the symmetry restrictions. Following this procedure, the drawback of the orbital variance is solved by defining a unique description for the starting orbitals and, consequently, for the INT-MP2-F12 method.

Figure 3.3 shows how the interference correction affects the convergence of the correlation energy of the methane molecule. The cc-pVXZ-F12 basis sets were chosen, where X = D,T,Q. As reference values, the explicitly-correlated CCSD(T)(F12) values in the same basis were used. Conventional CCSD(T) has a very slow convergence rate to the basis set limit. As it has already been mentioned, for an accurate description, extremely large basis sets should be used (at least of septuple-zeta quality) or extrapolated values from at least quintuple- and sextuple-zeta quality. [69] The extrapolated CCSD(T) value from cc-pV5Z and cc-pV6Z basis sets is at -0.24044 E_h, while the CCSD(T)/cc-pV6Z is at -0.23884 E_h. [116] Indeed, the CCSD(T)(F12)/cc-pVQZ-F12 value ($-0.23988\ E_h$) is already very close to the CBS of the method.

Like in the case of the helium atom, adding only the second-order F12 component to the CCSD(T) energy overestimates the correlation energy. Even if the relative difference from the CCSD(T)(F12) energies of the *same* basis set is smaller (i.e. the deviation between the conventional and the

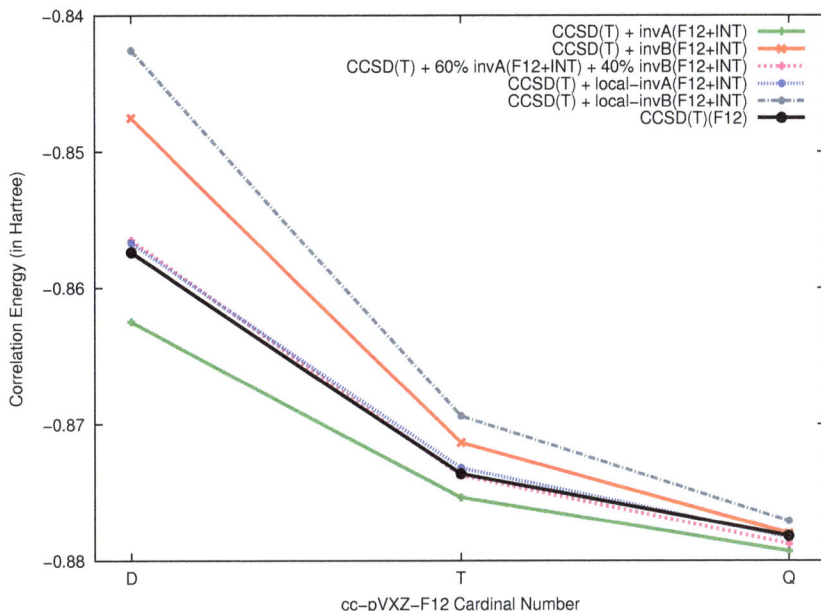

Figure 3.4: Conventional, corrected thought interference-corrected MP2-F12 and explicitly-correlated CCSD(T) correlation energy basis set convergence of the ozone molecule.

explicitly-correlated CCSD(T) becomes smaller when the second-order term is added), the basis set limit is still reached slowly. This behavior stems from the interference effect on the MP2-F12 energy terms. The *positive* δE_{INT} energy component brings the correlation energy of *each* basis set to the corresponding CCSD(T)(F12) energy. On Figure 3.3 results from invariant MP2-F12 with the approximation A and B are shown. Both approximations are reaching the basis set limit from above, with respect to CCSD(T)(F12), with approximation A being closer to the reference. Similar accuracy is obtained also with the application of the localization schemes. In particular, in Figure 3.3, results from Pipek-Mezey LMOs are shown while Boys LMOs also yield identical results.

The convergence behavior of the CCSD(T)-INT-F12 method is comparable with the one from the CCSD(T)(F12) level of theory: both converge faster to the basis set limit than the conventional coupled-cluster and their energies are almost identical. However, there is one important difference: the relative expensive explicitly-correlated part of the coupled-cluster method is substituted with the interference-corrected second-order F12 calculation. The δE_{INT} is a computational-free energy term which corrects the second-order F12 energy to converge to the CCSD(T) basis set truncation error. In the nest Chapter, timings between the two different methods will be given in order to highlight the essence of the interference effects.

3.3.4 Symmetry issues and proposed solutions

Unfortunately, INT does not perform always in the same manner like in the case of the methane molecule. There are specific cases in which CCSD(T)-INT-F12 should be used with caution. These cases can be divided in two groups. The first has to do with open-shell cases, like the correlation energy of atoms which are important for the calculation of the atomization energies and reaction barrier heights (Chapter 4). An efficient treatment to circumvent the problems arising for this cases is the orbital alignment which was described in Section 3.2.2 or the use of SCF orbitals obtained with symmetry restrictions. The second group has to do with molecules for which canonical orbitals fail to obtain the correct amount of correlation energy, by yielding a non-physical pair energy scaling. In some of these cases, symmetry restrictions on the SCF orbitals may solve this issue. However, this approach does not guarantee the calculation of the correct δE_{INT} term.

An illustrative example which shows such behavior is the ozone molecule. Figure 3.4 shows the convergence behavior of the correlation energy of this moiety. From this graph, the CCSD(T) energy convergence was excluded as far as it lies much higher (and thus, it converges too slow) from those of the CCSD(T)-INT-F12 and CCSD(T)(F12) methods. Therefore, attention has been given at the energy gap between -0.84 and -0.88 Hartree and to the importance of the second-order corrections δE_{INT} and δE_{F12}. On the contrary to the methane molecule (Figure 3.3), results from approximation F12/A are below the corresponding CCSD(T)(F12) energies (Figure 3.4). At the same time, approximation F12/B converges always from above. This different behavior may leads to inconsistency in the method and to large errors like in atomization or interaction energies. Therefore, extra consideration should be taken into account.

An attractive approach which can recover the correct convergence to the basis set limit is to sum the second-order energy terms from the two different approximations (A and B) with a weighted coefficient. This empirical scaling has been proposed from Samson and Klopper. [117] Based on their results, they used 60% of the F12/A and 40% of the F12/B to obtain a more accurate MP2-F12 correlation energy. By extending this idea, the $\delta E_{INT/A}$ and δE_{F12A} terms are weighted by 60% and those from approximation B by 40%. Indeed, this shifts the CCSD(T)-INT-F12 energy (pink line in Figure 3.4) exactly to the highly accurate CCSD(T)(F12) energies. However, there is a small deviation at the quadruple-zeta level. This perhaps is an indication that this empirical approach may work for small basis sets (double or triple-zeta quality) but it may introduce errors when it is used with larger basis.

A different procedure which solves the erroneous convergence of the CCSD(T) method with second-order corrections from interference effects of the ozone molecule, without falling back on empirical scaling, is the use of localization schemes. For the methane molecule (Figure 3.3) Boys LMOs did not affect (positively or negatively) the convergence behavior of the CCSD(T)-INT-F12 method. On the contrary, and as it is shown in Figure 3.4, the application of LMOs to the interference-corrected MP2-F12 leads to a convergence behavior similar to the one which was observed for the methane molecule: the CCSD(T) correlation energy with the terms from the INT-MP2-F12 method calculated with the invariant approximation A and with the Boys LMOs coincides with the CCSD(T)(F12) correlation energies.

Table 3.5: The energy levels (in a.u.) of the 9 valence orbitals of the ozone molecule.

Canonical	vs.	Localized
-0.489	9	-0.846
-0.553	8	-0.846
-0.565	7	-0.846
-0.783	6	-0.846
-0.801	5	-0.971
-0.834	4	-0.987
-1.094	3	-0.987
-1.433	2	-0.987
-1.752	1	-0.987

The explanation for this corrected behavior can be given by considering the effect of the canonical and the localized molecular orbitals to the e_{ij}^{INT} terms. It should also kept in mind that the wave function of a specific part of the molecule is described from small contributions of all orbitals. [118] The canonical orbitals are delocalized over the whole molecule. On the contrary, a single or a few localized orbitals would contribute the larger fraction at a given point while the rest can be neglected. Apart from the reduction of the computational time and disk storage introduced from the LMOs (linear scaling), they also recover the correct fraction of the interference correction (Figure 3.4). The energy levels of the occupied valence canonical and localized orbitals are shown in Table 3.5. For the canonical ones, every orbital lies in a different energy level. On the contrary, localized orbitals can be splitted in three subgroups: two subgroups of fourfold degenerate orbitals and one non-degenerate. Apart from their energy levels, the shape of the degenerate orbitals is equivalent. From these nine valence orbitals, the ij pair energies of the INT-MP2-F12 theory will be computed.

3.3.5 First results

A first step to understand the applicability of the interference effects on the CCSD(T) energy was the calculation on the atoms C, N, O and F, as well as on 16 small, closed-shell molecules. [41] These calculations were performed in the cc-pVXZ-F12 (X = D, T) basis sets of Peterson and co-workers [103], using their corresponding OptRI auxiliary basis as complementary auxiliary basis sets (CABS). [119] The aug-cc-pwCV(X + 1)Z cbas of Hättig [120] was used for the robust fitting of both the F12 integrals and the usual electron-repulsion integrals. The aug-cc-pV(X+1)Z jkbas basis of Weigend [121] was used for the two-electron contributions to the Fock matrix. The orbital invariant ansatz was applied with full optimization of the F12 amplitudes. On this specific study, only canonical orbitals obtained from SCF under symmetry restrictions were used.

Results with the cc-pVTZ-F12 basis for all atoms and molecules are shown in Table C.1 in Appendix C. For this basis set, the difference between the approximations F12/A and F12/B is not-negligible and as it has already discussed above, the convergence to the basis set limit, in respect to the CCSD(T)(F12), is either for both A and B from above (Figure 3.3) or not (Figure 3.4).

It has already been shown that the interference factors of the CBS model chemistries can be used as a recipe for estimating the basis set truncation errors of high-level methods from the MP2 errors. [7,41] For this, the interference corrections are significant. They constitute about one third in magnitude of the MP2-F12/A and MP2-F12/B estimates of the basis set truncation error. This can be shown by comparing the second-order basis set truncation error (Eq. (2.23)) with the higher-order (CCSD(T)) basis set truncation error (Eq. (2.24)). Figure 3.5 shows that the sums $\delta E_{INT/A} + \delta E_{F12/A}$ (■ symbols) and $\delta E_{INT/B} + \delta E_{F12/B}$ (○ symbols) are in much better agreement with the F12 contribution to the coupled-cluster energy than the plain $\delta E_{F12/A}$ and $\delta E_{F12/B}$. The F12 contribution to the coupled-cluster energy is computed as $\delta E_{CCSD(T)(F12)} - \delta E_{CCSD(T)}$. In particular, the $\delta E_{INT/A} + \delta E_{F12/A}$ data lie almost perfectly on the ideal line, showing the one-to-one correlation between the second- and higher-order errors. This is a clear verification of the significance of the interference factor and the applicability of the proposed method. Only the values for nitrous oxide (N_2O) and ozone (O_3) are located slightly below this line, that is, overestimate the $\delta E_{CCSD(T)(F12)} - \delta E_{CCSD(T)}$ difference slightly in magnitude. However, what should be noted is that these energies have been obtained with (semi-)canonical orbitals. Therefore, for cases such as O_3, this may lead to deviations from the CCSD(T)(F12) energies, as it was discussed in Section 3.3.4.

Figure 3.6 illustrates the interference correction to the MP2-F12 estimate of the basis set truncation error when considering atomization energies, i.e. the electronic energy required to fully

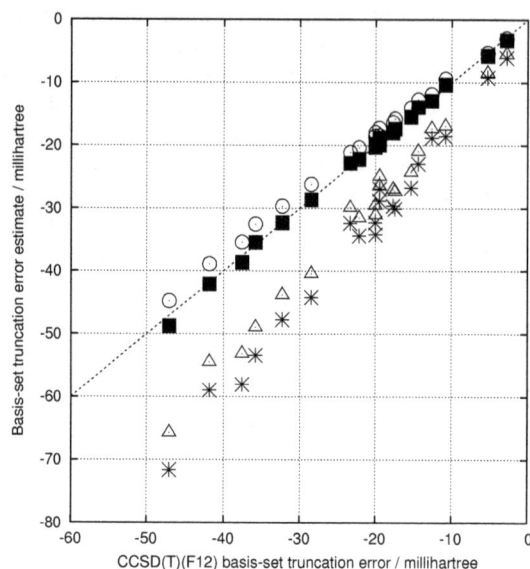

Figure 3.5: Basis set truncation error (in mE_h) of the total energy computed at the MP2-F12/A level, with (■) and without (∗) interference correction, plotted against the basis set truncation error (in mE_h) computed at the CCSD(T)(F12) level. Results for the MP2-F12/B are also given, with (∘) and without (△) interference correction.

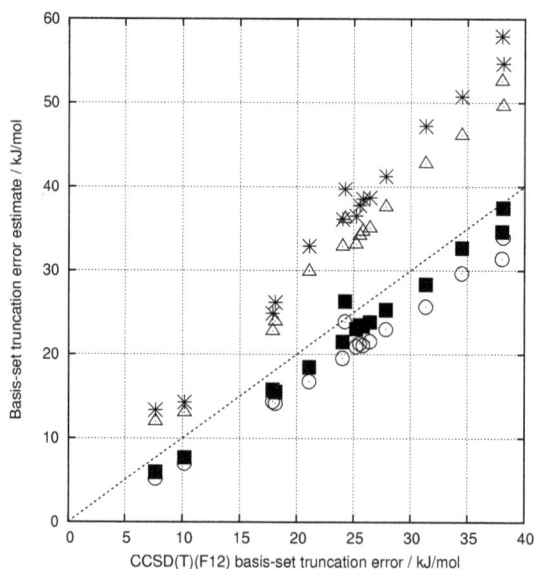

Figure 3.6: Basis set truncation error (in kJ/mol) of the atomization energy computed at the MP2-F12/A level, with (■) and without (∗) interference correction, plotted against the basis set truncation error (in kJ/mol) computed at the CCSD(T)(F12) level. Results for the MP2-F12/B are also given, with (○) and without (△) interference correction.

Table 3.6: Correlation energy components of interference-corrected MP2-F12 (in E_h) of the neon atom and the neon dimer ($d_{Ne\cdots Ne} = 50$ bohr). Interference corrections from different starting orbitals (canonical, Boys, Pipek-Mezey) are also included. The YP-aug-cc-pVTZ basis has been used.

	δE_{CCSD}	$\delta E_{CCSD(T)}$	$\delta E^{canonical}_{F12/A}$	$\delta E^{canonical}_{INT/A}$	$\delta E^{canonical}_{F12/B}$	$\delta E^{canonical}_{INT/B}$
Ne atom	$-0.274\ 093\ 2$	$-0.279\ 373\ 1$	$-0.045\ 670\ 5$	$0.010\ 711\ 5$	$-0.041\ 020\ 5$	$0.009\ 672\ 9$
Ne dimer	$-0.548\ 186\ 3$	$-0.558\ 746\ 2$	$-0.091\ 340\ 9$	$0.021\ 422\ 9$	$-0.082\ 040\ 9$	$0.019\ 345\ 7$
Difference	0	0	0	0	0	0

			$\delta E^{boys}_{F12/A}$	$\delta E^{boys}_{INT/A}$	$\delta E^{pipek}_{F12/A}$	$\delta E^{pipek}_{INT/A}$
Ne atom			$-0.045\ 670\ 5$	$0.010\ 947\ 1$	$-0.041\ 020\ 5$	$0.010\ 711\ 5$
Ne dimer			$-0.091\ 340\ 9$	$0.021\ 894\ 3$	$-0.082\ 040\ 9$	$0.021\ 422\ 9$
Difference			0	0	0	0

dissociate the molecule into its constituent atoms. Obviously, the plain MP2-F12 estimates for the basis set truncation error are significantly improved by adding the interference corrections. However, the agreement with coupled-cluster theory is not perfect. In particular, the $E_{INT/A} + E_{F12/A}$ estimates appear to be systematically too small. In Chapter 4, possible solutions to this issue will be presented.

3.3.6 Size-extensivity

An important feature of the CCSD(T)-INT-F12 method is its size-extensivity. On Table 3.6 a numerical study is shown. The neon dimer has been chosen, with an internuclear distance at 50 bohr. Calculations with the YP-aug-cc-pVTZ basis [119] and with different starting orbitals (canonical, Boys and Pipek-Mezey) have been done. As it is well known, coupled-cluster and explicitly-correlated methods are size-extensive and thus, main concern are the interference terms. In all cases, the energy components of the stretched geometry are identical with the double amount of the respective term of the neon atom. An interesting observation is that the $E^{pipek}_{INT/A}$ energies obtained from the Pipek-Mezey localization scheme are equal to the ones from the canonical orbitals (obtained from SCF under D_{2h} symmetry restrictions), while the corresponding INT term from the Boys localization differs by about 0.5 mE_h. However, for the stretched geometry, relative energies from all approaches predict that there is no interaction between the two neon atoms. A more detailed discussion about weak interactions and the potential curves of the helium, neon and argon dimers will be given in Chapter 5.

3.3.7 Interference-corrected two-electron Darwin term

Finally, it is worth to mention a recent application of the interference-correction which is in a similar context with the one described in this Chapter. The MP2 basis set truncation error of the two-electron Darwin term (D2), a term related with the first-order relativistic corrections to the Hamiltonian [14], was scaled down through the interference effect in order to obtain a good estimate of the corresponding coupled-cluster basis set truncation error. [122] Even if the D2 term has a different convergence behavior ($\propto X^{-1}$) than the correlation energy ($\propto X^{-3}$), interference corrected explicitly correlated perturbation theory performs in a very favorable way: it provides a very accurate estimate of the CCSD basis set truncation error ($E_{CCSD(F12)/2B} - E_{CCSD}$).

Additivity Scheme

for SB

    obtain SCF canonical orbitals with **symmetry** or use **localization** schemes

    build coefficient matrix $\mathbf{A}_{SB}$

    write e_{ij}^{MP2} pair energies and interference factors in file

    calculate $\delta E_{CCSD(T)}$ correlation energy

for LB

    obtain SCF canonical orbitals with **symmetry** or use **localization** schemes

    build coefficient matrix $\mathbf{A}_{LB}$

    build overlap matrix $\mathbf{S}$ between the two basis sets

    calculate transformation matrix $\mathbf{F}$ for **orbital alignment**

    if ($\mathbf{F}$ not diagonal) **then**

        do while $\mathbf{F}$ not diagonal

            re-order non-diagonal orbitals

        end do

    end if

    write e_{ij}^{F12} pair energies in file

calculate e_{ij}^{INT} pair energies

add $\delta E_{CCSD(T)}$ to 2nd-order energy corrections   (Eq. (3.22))

add $E_{HF/LB}$ to correlation energy

Algorithm 2: Pseudocode of the additivity scheme.

3.4 Additivity Scheme

In the previous sections, the faster convergence to the basis set limit of CCSD(T) through the second-order interference effect has been shown. However, the computationally most expensive part of the method still remains the time needed for the CCSD(T) calculation. For surpassing this problem, an alternative approach was considered, in which different basis sets are used for different levels of theory. This dual-level approach [123] has been used successfully in the composite schemes discussed in Subsection 2.2.4, like the HEAT and ccCA protocols. A similar idea was used in a recent study on atomization energies in which, various corrections were added at the CCSD(T) level. [114] One of these corrections was about the second-order truncation error

$$\delta E_{F12} = f_{int}\left(E_{fc\text{-}MP2\text{-}F12} - E_{fc\text{-}MP2/cc\text{-}pCVQZ}\right). \tag{3.20}$$

In this study, an empirical fixed interference factor $f_{int} = 0.78$ was tested, optimized by minimization of the mean deviation for the test set used, and different basis sets were used for different levels of theory.

This approach has also been used successfully for obtaining accurate energies between weakly interacting systems and it is based on the consideration that the difference between the CCSD(T) and MP2 (interaction) energies depends only negligibly on the basis set size and thus, it can be determined with small or medium size basis sets

$$E_{CCSD(T)/CBS} \approx E_{MP2/CBS} + \left(\delta E_{CCSD(T)} - \delta E_{MP2}\right)_{\text{small basis}} \tag{3.21}$$

This assumption has been used successfully in various studies about systems which are stabilized with van der Waals forces. [11, 124–126] Further discussion about this kind of applications will be

---

given in Chapter 5.

Having as starting point the above considerations, the effectiveness of the interference correction on the second-order terms ($E_{MP2/CBS} - E_{MP2/SB}$, where SB stands for small basis set) was examined. This is done explicitly for each pair energy difference and by re-ordering the energy terms of Eq. (3.21), the equation which constitutes the *additivity scheme* is obtained

$$\delta E_{CCSD(T)/CBS} \approx \delta E_{CCSD(T)/SB} + \sum_{ij} F_{SB}^{ij} [e_{MP2-F12/LB}^{ij} - e_{MP2/SB}^{ij}], \tag{3.22}$$

where LB stands for a relatively larger basis set than the SB. A simple program was written which controls the calculations with the different basis sets and prints the interference-corrected second-order and total correlation energies (Algorithm 2). The extrapolation of the CCSD(T) energy to the basis set limit is done with the SB and thus, the interference factors F^{ij} are obtained from that basis set.

After obtaining SCF canonical orbitals with symmetry restrictions with the SB set, the MP2 pair energies and the corresponding interference factors for the ij pairs are stored on disk. The CCSD(T) correlation energy is then calculated.

As a second step, the SCF canonical orbitals with symmetry restrictions for the LB set are obtained. However, for degenerate or nearly degenerate orbitals, the sequence of the MOs may not be identical between calculations with different basis sets. This causes no problems to the total energy when one individual basis set is used but it leads to errors when different ij spin-orbital pairs are compared, like in Eq. (3.22). In order to avoid this error, a transformation matrix $\mathbf{F}$ was built which checks the relative *orbital alignment* between the two different basis sets. It is constructed from the MO coefficients of the two different basis sets ($^{\kappa}\mathbf{A}_{SB}$ and $^{\kappa}\mathbf{A}_{LB}$) and the overlap matrix between SB/LB sets

$$\mathbf{F} = {}^{\kappa}\mathbf{A}_{LB}^{\dagger} \, \mathbf{S} \, {}^{\kappa}\mathbf{A}_{SB}, \tag{3.23}$$

where κ corresponds either to the MOs (restricted cases) or to the α and β spin-orbitals (unrestricted cases). All the matrices involved are constructed from the coefficients in spherical harmonic orbitals. If $\mathbf{F}$ is not in diagonal form, then the sequence of the orbitals is not identical between the two different basis sets. Alignment of the MOs is achieved by re-ordering the orbitals and re-calculating the transformation matrix. After one or two rotations, the transformation matrix is on the desired diagonal form. Then, the MP2-F12/LB calculation starts and the pair energies are stored on disk. Summation over all pair energies based on Eq. (3.22) and addition to the CCSD(T)/SB energy gives the approximate CCSD(T)/CBS correlation energy.

An example about the error introduced from the non-aligned MOs is shown on Table 3.7. Results from the additivity scheme for carbonyl fluoride (CF_2O) have a big deviation when the double-zeta quality cc-pVDZ-F12 basis was used as LB and cc-pVTZ as SB. The triplet pair energies between the 10th and 15th MOs ($10\alpha\,15\alpha$ or $10\beta\,15\beta$ electron pairs) have a relative small difference for this basis set combination (-0.42496 mE_h for cc-pVTZ, -0.41051 mE_h for cc-pVDZ-F12 basis set). Same holds also for the $11\alpha\,15\alpha$ and $12\alpha\,15\alpha$ pairs, but not for the next two pairs. It is obvious that the pair energy of $13\alpha\,15\alpha$ from the cc-pVTZ basis corresponds to the $14\alpha\,15\alpha$ pair from cc-pVDZ-F12 and vice versa; the difference between -1.20889 and -5.90270 mE_h is large. The same behavior was observed for all pairs which included electrons from the 13th and 14th orbitals. Thus, rotation of these two MOs before the MP2-F12 calculation solves this problem. The energy difference between the calculation with and without the orbital alignment is 24.96 kJ/mol, which

Table 3.7: MP2 pair energies (in mE_h) and interference factor of carbonyl fluoride (CF_2O). cc-pVTZ is used as SB, cc-pVDZ-F12 as LB.

Electron Pair	cc-pVTZ	cc-pVDZ-F12	$F^{ij}_{cc-pVTZ}$
⋮			
$10\alpha\,15\alpha$	-0.42496	-0.41051	0.94579
$11\alpha\,15\alpha$	-2.86765	-2.81662	0.84152
$12\alpha\,15\alpha$	-0.18856	-0.18480	0.96419
$13\alpha\,15\alpha$	**-1.20889**	**-5.90270**	0.89787
$14\alpha\,15\alpha$	**-6.20355**	**-1.16488**	0.79651
⋮			

consists a significant amount of correlation energy.

The above problem appeared mainly (but not only) when double-zeta quality basis sets were used, either as SB or LB. It should also be mentioned at this point that during the search for the combination of the optimum basis sets for the additivity scheme, as "small" basis sets were also used relative large basis sets like, for example, the cc-pV5Z basis set. In this case, the SB has much more atomic basis functions from the "large" one but the nomenclature of SB and LB was kept, in sake of consistency. This was done in order to distinguish between the basis sets used for the CCSD(T) energy, the interference factor and the MP2 pair energies from the basis sets used for the MP2-F12 pair energies. Numerical results between the one-basis and the additivity scheme will be discussed in the next two chapters.

CHAPTER 4

Thermochemistry

In this chapter, the accuracy of the coupled-cluster singles, doubles and perturbative triples with corrections from the interference-corrected explicitly-correlated second-order perturbation theory for atomization energies and reaction barriers is being examined. Atomization energy is the energy required to fully dissociate the molecule into its constituent atoms and is being calculated as the difference between the energy of the atoms and the molecule itself. Since atomization energies represent energy differences in the energies of systems containing different numbers of paired electrons, they pose a challenge for pair-electron theories and thus, for the interference-corrected MP2. They also require a high degree of flexibility in the description of the short-range interactions. [14] On the contrary, a reaction barrier, the energy barrier to chemical reaction, usually can be computed with higher accuracy. It consists the energy difference between the energies of the transition state and the reactants.

Various test sets are examined with different basis sets: the 105-molecule test set plus the H_2 molecule (106-molecule test set) of Bakowies [33], the heats of formation with respect to H_2, CO, CO_2, F_2 and N_2 of 25 molecules from the 106-molecule test set, the six-membered atomization energy (AE6) and the six-membered reaction barrier height (BH6) test sets of Lynch and Truhlar [127] and the G2/97 test set [128,129] which was discussed in Chapter 2.2.4. Statistical results showing the accuracy of the method will be shown and discussed. Target was not to reach the level of chemical accuracy (i.e. less than 1 kcal/mol), but the stricter sub-chemical accuracy, which is less than 1 kJ/mol.

4.1 106-Molecule Test Set

4.1.1 Reference values

In 2007, Dirk Bakowies [33] used a collection of 105 molecules containing H, C, N, O and F atoms in order to show the applicability of his extrapolation scheme based on the deviations of the asymptotic $\sim X^{-3}$ convergence. He used a trial function X^β which, for an effective exponent β, provides the correct energy at the basis set limit.

The specific test set comprises an extensive collection of molecules which are composed of H, C, N, O and F atoms. All species of the set have maximum six atoms and this fact makes it suitable for direct testing new methods and schemes. Additionally, it does not exclude "difficult" cases, like molecules with multi-reference character (e.g. O_3, N_2O_3, cyclobutadiene), or species which have not an experimental atomization energy archived in the Active Thermochemical Tables (ATcT) database [130,131], like the linear dicyanoacetylene (C_4N_2) or the tetrahedral carbon cluster named tetrahedran (C_4H_4). These cases make the choice of the 106-molecule test set a challenge for the newly introduced interference-corrected method.

The same test set (plus the dihydrogen molecule) was used by Klopper and co-workers [114, 132]. In the first article [114], the atomization energies of the 106 molecules were computed with an estimated standard deviation (from the values compiled in the ATcT) of ± 0.13 kJ/mol per valence electron in the molecule. This was achieved by means of a composite scheme which included a series of contributions: the correction to the basis set truncation error obtained at the MP2-F12 level, the correction for anharmonicity effects and zero-point vibrational energy, relativistic correction, the correction accounting for the improvement between the CCSD(T) and CCSDT coupled-cluster models, and the correction for the perturbative treatment of the connected quadruple excitations (Q). It is worth pointing out that in this study, a second-order correction of the form of Eq. (3.20) was added. This was based on an empirical fixed value for the interference factor ($f_{int} = 0.78$) and it provided an excellent accuracy with the 73 experimental atomization energies of the ATcT tables: a mean absolute deviation of 0.90 kJ/mol and a root-mean-square (RMS) of 1.22 kJ/mol is a clear proof of the accuracy of the proposed composite scheme.

The second article [132] was mainly focused on obtaining high-accurate CCSD energies from the explicitly-correlated coupled-cluster code of TURBOMOLE [133]. The frozen-core (fc) CCSD(F12) with both the invariant and fixed ansätze were used to obtain the basis set limit of the coupled-cluster model, while the basis set limit of the perturbative triples were estimated from a two-point extrapolation. One extra energy component was also added to the total atomization energies in order to improve the Hartree-Fock part. This was computed from the difference between the HF energy from the basis set which was used for the CCSD(F12) calculation (cc-pCVQZ) and the HF energy from a basis set with a cardinal number larger by one (cc-pCV5Z).

The main difference between these two articles is the approach from which the basis set limit of the coupled-cluster singles-and-doubles is being computed. In the first article [114], the CCSD limit is estimated from corrections which have been calculated from the empirically scaled MP2-F12. In the second work [132], the frozen-core explicitly-correlated CCSD method is used for the calculation of the basis set limit. The choice of a smaller basis set (def2-QZVPP) used for the CCSD(F12) energy, instead of the quintuple-zeta quality basis (cc-pCV5Z) used in the first article for the conventional CCSD, leads to a considerably less time consuming composite scheme, without significant loss of accuracy. Comparison of the computational results with the experimental atomization energies for the 73 out of 106 molecules of the test set provides a clear picture of the accuracy which can be achieved from these two approaches. Conventional CCSD(T) with second-order corrections, scaled with the empirical factor ($f_{int} = 0.78$), has a RMS deviation of 1.22 kJ/mol. The CCSD(F12) with either the fixed or invariant ansatz, which include an extrapolated (T) energy component, yield results with similar accuracy. Both explicitly-correlated approaches have RMS deviations below 2 kJ/mol: 1.7 kJ/mol for the invariant CCSD(T)(F12) method and 1.5 kJ/mol for the fixed.

An interesting conclusion was drawn from these two articles. Reference [132] fairly ends with the conclusion that "the empirical factor of 0.78 chosen to scale the MP2-F12 corrections was very reasonable in view of the interference factor of Petersson's CBS theory". In the present Chapter the above statement is discussed, with the main difference that the effect of an interference factor *explicitly* calculated for each pair of electrons is under consideration.

Apart from the aforementioned considerations, the target of these publications was to provide an accurate and reliable ground for an additivity scheme based on the CCSD(T)(F12)/def2-QZVPP level. This target partially also applies to the present study. In sake of formulating a less computational demanding approach, without conventional coupled-cluster with a large basis or explicitly-correlated coupled-cluster with middle-size basis (e.g. of quadruple-zeta quality), results from CCSD(T) with second-order corrections from INT-MP2-F12 are being compared with high-accurate CCSD(T)(F12) correlation energies. The savings from the new method have mainly to do with computational time and hard disk space requirements. The geometries used in the previous two studies were also employed here. The molecules have been obtained from optimizations at the all-electron (ae) CCSD(T)/cc-pCVTZ level (cc-pVTZ for hydrogen). The reference correlation component of the atomization energies have been calculated as

$$\delta E^{\text{ref}} = \delta E_{\text{CCSD}} + \delta E_{(\text{T})} + \delta E_{\text{F12}}. \tag{4.1}$$

The first and third terms have been computed from fc-CCSD(F12) calculations with the def2-QZVPP basis [134]. The $\delta E_{(\text{T})}$ energy has been evaluated from a two-point extrapolation at the fc-CCSD(T) level with the cc-pCVQZ and cc-pCV5Z basis sets [22].

As a first step, the correlation energy obtained from the newly proposed method will be compared with the one from Eq. (4.1). Secondly, the total atomization energy will be discussed, with the addition of extra energy correction terms. Moreover, results of the new method will be compared with the 73 (out of 106) experimental available atomization energies of the ATcT tables and with high-level calculations from Ref. [114]. In that study, the total electronic energy was computed from the fc-CCSD(T) method with the cc-pCVQZ basis set. A series of corrections have been added to the total energy, as it was discussed above, including a second-order correction from MP2-F12 with a fixed interference factor at 0.78. This composite scheme has a RMS of 1.22 kJ/mol over the 73 molecules with an experimental value at the ATcT tables. For obtaining accuracy compared with these high-level calculations and experimental values, similar corrections to the electronic total energy should be taken into account. These include higher-order corrections, like for the connected triple excitations, calculated as the difference between CCSDT/cc-pVTZ and CCSD(T)/cc-pVTZ correlation energies, and a perturbative correction for connected quadruple excitations (CCSDT(Q)), calculated as the difference between CCSDT(Q)/cc-pVDZ and CCSDT/cc-pVDZ correlation energies. In addition, scalar relativistic effects, spin-orbit interactions for the atoms, and anharmonic zero-point vibrational energies were included. All these data have been taken from Ref. [114] and are added to the CCSD(T)-INT-F12 energies. In the next Sections they will be mentioned as "Other". One extra correction which was recalculated for the current study is the correction for core/core-valence correlation effects. This energy component was obtained from the difference between the coupled-cluster correlation energy from an all-electron and a frozen-core calculation

$$\delta E_{\text{CV}} = \delta E_{\text{ae-CCSD(T)/cc-pCVQZ-F12}} - \delta E_{\text{fc-CCSD(T)/cc-pCVQZ-F12}}. \tag{4.2}$$

The quadruple-zeta quality correlation consistent basis sets which have been optimized for accurately describing core-core and core-valence correlation effects with explicitly correlated F12 methods (cc-pCVQZ-F12) have been used. [135] Finally, the first-order correction to the HF energy obtained from the explicitly-correlated MP2-F12 method (CABS singles) is taken also into account. CABS singles are discussed in detail in Chapter 4.1.4. All energy contributions added in the current composite scheme are shown in Eq. (4.3)

$$\begin{aligned} E_{\text{Composite Scheme}} = \quad & E_{\text{HF}} + \delta E_{\text{CABS Singles}} + \\ & \boldsymbol{\delta E_{\text{CCSD(T)}}} + \boldsymbol{\delta E_{\text{INT}}} + \boldsymbol{\delta E_{\text{F12}}} + \\ & \delta E_{\text{CV}} + \delta E_{\text{T}} + \delta E_{(\text{Q})} + \\ & \delta E_{\text{ZPVE}} + \delta E_{\text{Anh}} + \delta E_{\text{MVD}} + \delta E_{\text{SO}}. \end{aligned} \tag{4.3}$$

The correlation energy components of the CCSD(T)-INT-F12 method are shown with bold font.

4.1.2 Computational details

All calculations have been performed with the TURBOMOLE program package. For the conventional and explicitly-correlated coupled-cluster calculations the 6.2 version was used. Interference-corrected MP2-F12 calculations have been performed with a local version of TURBOMOLE (see Algorithm 1 from Chapter 3.2).

The equilibrium geometries of all 106 molecules were obtained from the previous studies [114, 132] at the all electron CCSD(T) level in the correlation-consistent core-valence triple-ζ basis set (cc-pCVTZ) of Woon and Dunning [19, 22]. For the hydrogens contained in these molecules, the cc-pVTZ basis was used. All the molecules are closed-shell species and therefore, the RHF wave function was used as the reference for the geometry relaxation. The molecules included in the specific test set are listed in Table 4.6.

Different basis sets or combinations of basis sets for the additivity scheme (see Section 3.4) were used for the current study in order to establish an optimum protocol which can reach high accuracy for the atomization energies of the 106-molecule test set. Two different families of basis sets were under consideration: the cc-pVXZ basis $(X = D, T, Q, 5)$ of Dunning [19] and the cc-pVXZ-F12 basis $(X = D, T, Q)$ of Peterson and co-workers [103]. Both families have been developed within a correlation consistent methodology, with the second being specifically optimized for explicitly-correlated calculations. Both families were used for the "one-basis" approach, i.e. all energy terms of the CCSD(T)-INT-F12 method are obtained with the same basis. For the additivity scheme, the cc-pVXZ family of basis set was used as "small basis" (SB) and the relatively larger cc-pVXZ-F12 family as "large basis" (LB). It should be reminded that the performance of the latter family within the INT-MP2-F12 framework has already been discussed in Section 3.3 and in Ref. [41].

Dependent on the choice of the basis set family which was used for the interference-corrected MP2-F12, the corresponding cc-pVXZ-F12 cabs auxiliary basis have been used. The aug-cc-pwCV$(X + 1)$Z cbas of Hättig [120] (aug-cc-pV$(X + 1)$Z for the hydrogens) was used for the robust fitting of both the F12 integrals and the usual electron-repulsion integrals. The aug-cc-pV$(X + 1)$Z jkbas basis of Weigend [121] was used for the two-electron contributions to the Fock matrix. The same auxiliary basis (cbas) were used for the calculations with the cc-pCVQZ-F12 basis for the core/core-valence correction.

The orbital invariant ansatz with full optimization of the amplitudes was used in all explicitly-correlated calculations (CC and MP2), including "spin-flip" amplitudes [136–138] for open-shell atoms, with the ansatz 2 and the F + K approximation [102] for the commutator of kinetic energy with Slater-type geminal $f_{12} = \gamma^{-1}\{1 - \exp(-\gamma r_{12})\}$. The exponents of γ of the latter expression are different for every basis set. For the cc-pVXZ-F12 sets, they were taken as recommended by Peterson and co-workers: $\gamma = 0.9\,a_0^{-1}$ for cc-pVDZ-F12, $\gamma = 1.0\,a_0^{-1}$ for cc-pVTZ-F12 and $\gamma = 1.1\,a_0^{-1}$ for cc-pVQZ-F12. For the cc-pVXZ basis, the default global exponent $\gamma = 1.4\,a_0^{-1}$ was used for all the members of the cc-pVXZ series.

All calculations have been performed in the frozen-core (fc) approximation, except from those for the core/core-valence corrections. Unrestricted Hartree-Fock (UHF) and restricted open-shell Hartree-Fock (ROHF) references were used for the open-shell atoms and restricted HF references for the closed-shell molecules. In both cases, HF calculations have been done under symmetry restrictions, for reasons that have been discussed in Chapter 3.3.

4.1.3 (T*) correction

CCSD(T) with second-order corrections from interference-corrected MP2-F12 approximates the CCSD(T)(F12) method, as it was discussed in Section 3.3. In the current implementation of the explicitly-correlated CCSD(T) method, the F12 geminals are included only for the coupled-cluster singles and doubles. This means that the basis set limit of CCSD is obtained and the perturbative contributions of triple excitations are being added from the conventional CCSD(T). On the contrary, the reference values with which the INT-MP2-F12 results will be compared, include extrapolated $\delta E_{(T)}$ energy terms from a two-point fit [132]. The higher-order extrapolated (T) energy component is missing in the new method. For that reason, an empirical scaling of the perturbative triples energy term was examined, as it has been proposed by Marchetti and Werner [139]

$$\delta E_{(T^*)} = \delta E_{(T)} \cdot \frac{\delta E_{\text{MP2-F12}}}{\delta E_{\text{MP2}}} \tag{4.4}$$

The nominator ($E_{\text{MP2-F12}}$) of the scaling factor of the perturbative triples energy is always larger than the denominator, as far as the explicitly-correlated MP2 reaches the basis set limit of the method and is always lower than the conventional MP2. This empirical scaling factor has typically values between 1.05 and 1.15. Assuming that the fraction of the correlation energy obtained with the given basis is the same for the MP2 correlation energy and the (T) contribution, a rough estimate of the basis set limit of the (T) energy term can be obtained.

For the additivity scheme results, the $E_{(T)}$ of the small basis (SB) is scaled from the ratio of the MP2-F12 energy obtained from the large basis (LB) and the MP2 energy from the SB. This compromise was made from the fact that the explicitly-correlated MP2 energy with a more complete basis reaches faster the CBS of the method.

The empirical (T*) correction has been applied successfully for weakly interacting systems with predominant dispersion contributions, hydrogen bonded systems or complexes with mixed contributions to the interaction energy. [139] For counterpoise corrected interaction energies, the ratio $E_{\text{MP2-F12}}/E_{\text{MP2}}$ is slightly different for the monomers and the dimer, and this leads to a size-consistency error. The authors had suggested the use of the same factor for each molecule. They determined the ratio for the dimer and they used it for the monomers as well. (Some other authors [140] name this size-consistent approach as (T**), while keeping the notation (T*) for the individual scaling of the molecule and the atoms or the dimer and the monomers. This notation will also be followed in the next sections.) For a small test set of eleven weakly interacting complexes which are dominated by dispersion forces [139], the differences of the CCSD(T*)-F12a results from their reference values did not exceed 0.2 kcal/mol. However, CCSD(T*)-F12a results have been compared with extrapolated CCSD(T) energies using a $E_n = E_{\text{CBS}} + An^{-3}$ formula, where n is the cardinal number of the basis set and A are determined by fitting to the energies for $n = 3$ and $n = 4$. The choice of triple- and quadruple-zeta basis sets is insufficient to provide the real CBS limit, especially for more demanding applications like atomization energies and reaction barrier heights, and thus, the use of empirical (T*) can also be non-transferable to these applications. Further discussion on the effectiveness of the (T*) term will be given in the next Sections.

4.1.4 One-basis scheme

The discussion on the applicability of the CCSD(T)-INT-F12 method for obtaining accurate atomization energies will be started by focusing on the correlation energy of the 106 molecules. But

Table 4.1: Statistics of the deviations (in kJ/mol) of the computed correlation energy values calculated with the conventional CCSD(T).

Basis Set	Mean Error	MAD	RMS	Max Error	Molecule
cc-pVDZ	134.32	134.32	141.75	245.14	72.C_4N_2
cc-pVDZ-F12	102.34	102.34	108.37	189.74	72.C_4N_2
cc-pVTZ	54.53	54.53	57.58	113.98	104.N_2O_4
cc-pVTZ-F12	42.65	42.65	44.94	82.28	104.N_2O_4
cc-pVQZ	21.52	21.52	22.84	49.41	104.N_2O_4
cc-pVQZ-F12	17.91	17.91	18.92	35.93	104.N_2O_4
cc-pV5Z	10.19	10.19	10.89	24.74	104.N_2O_4

Table 4.2: Statistics of the deviations (in kJ/mol) of the computed correlation energy values calculated with the CCSD(T) method plus second order corrections from the MP2-F12/A.

Basis Set	Mean Error	MAD	RMS	Max Error	Molecule
cc-pVDZ	-26.85	26.99	29.99	-78.85	104.N_2O_4
cc-pVDZ-F12	-16.89	16.89	17.94	-35.50	104.N_2O_4
cc-pVTZ	-13.11	13.11	14.03	-29.37	104.N_2O_4
cc-pVTZ-F12	-13.13	13.13	13.91	-24.49	66.C_3H_8
cc-pVQZ	-9.11	9.11	9.64	-18.28	72.C_4N_2
cc-pVQZ-F12	-8.92	8.92	9.45	-15.64	72.C_4N_2
cc-pV5Z	-5.95	5.95	6.29	-12.51	72.C_4N_2

before analyzing the results of the new method presented in this Thesis, a brief comment should be given for the correlation energy of the conventional CCSD(T) method. In Table 4.1 the statistical results of the uncorrected CCSD(T) method are shown. The deviations from the reference values are significantly large, even for the largest basis under consideration in this study. The mean absolute deviation of 10.19 kJ/mol for the quintuple-zeta quality basis highlights the need of moving beyond the quantum chemistry's "golden standard". This result was expected since the CCSD(T) method needs extremely large basis sets (septuple-zeta or larger) in order to reach the basis set limit of the method and achieve higher accuracy. As it has already been discussed in Chapter 2, this can be accomplished either with the use of even larger basis sets or the addition of energy corrections based on extrapolation schemes which lead to a faster basis set convergence.

The direction of the current study is the use of corrections obtained from interference effects on the pair energies of the explicitly-correlated MP2 method. Firstly, the statistical results of a scheme where the interference effects are neglected are shown in Table 4.2. This means that only the δE_{F12} energy component is added to the CCSD(T) energy or, in other words, the interference factors of the pair energies are set to one. This consideration is similar with those of ref. [114]. For all basis sets, the MAD and the RMS values are significantly lower. The RMS of the double-zeta quality basis are reduced by a factor of five, for the cc-pVTZ basis by a factor of four and for the rest at around a factor of two. The negative values of the mean errors indicate that the addition of only the δE_{F12} component overestimates the correlation part of the atomization energies, while lack of it significantly underestimates them. The CCSD(T) atomization energy approaches always the basis set limit from above, in respect with the reference CCSD(T)(F12)/def2-QZVPP level of theory.

Although the important reduction of the differences from the reference values that the second-

Table 4.3: Statistics of the deviations (in kJ/mol) of the computed correlation energy values calculated with the CCSD(T) method plus second order corrections from the interference-corrected MP2-F12/A.

Basis Set	Mean Error	MAD	RMS	Max Error	Molecule
cc-pVDZ	15.86	17.00	18.70	37.02	$98.H_3NO$
cc-pVDZ-F12	16.93	16.93	18.03	33.74	$70.C_4H_4$
cc-pVTZ	7.73	7.85	8.46	16.02	$98.H_3NO$
cc-pVTZ-F12	4.90	4.90	5.22	11.85	$104.N_2O_4$
cc-pVQZ	1.32	1.50	1.81	4.56	$104.N_2O_4$
cc-pVQZ-F12	0.46	0.66	0.84	3.12	$104.N_2O_4$
cc-pV5Z	-0.20	0.49	0.64	-2.19	$39.C_2HF_3$

Table 4.4: Statistics of the deviations (in kJ/mol) of the computed correlation energy values calculated with the CCSD(T*) method plus second order corrections from the interference-corrected MP2-F12/A.

Basis Set	Mean Error	MAD	RMS	Max Error	Molecule
cc-pVDZ	1.34	11.52	14.72	-45.07	$104.N_2O_4$
cc-pVDZ-F12	5.80	7.70	8.93	19.90	$70.C_4H_4$
cc-pVTZ	1.49	3.92	4.78	-12.38	$39.C_2HF_3$
cc-pVTZ-F12	0.20	1.61	2.11	-8.05	$103.N_2O_3$
cc-pVQZ	-1.49	1.85	2.40	-7.59	$103.N_2O_3$
cc-pVQZ-F12	-1.83	1.84	2.20	-6.56	$103.N_2O_3$
cc-pV5Z	-1.68	1.70	1.97	-5.31	$103.N_2O_3$

Table 4.5: Statistics of the deviations (in kJ/mol) of the computed correlation energy values calculated with the CCSD(T**) method plus second order corrections from the interference-corrected MP2-F12/A.

Basis Set	Mean Error	MAD	RMS	Max Error	Molecule
cc-pVDZ	0.50	11.57	14.93	-47.55	$104.N_2O_4$
cc-pVDZ-F12	4.96	7.53	8.81	19.78	$70.C_4H_4$
cc-pVTZ	0.94	3.88	4.82	-13.15	$39.C_2HF_3$
cc-pVTZ-F12	-0.07	1.72	2.31	-9.03	$103.N_2O_3$
cc-pVQZ	-1.77	2.07	2.65	-8.21	$103.N_2O_3$
cc-pVQZ-F12	-1.97	1.99	2.38	-7.05	$103.N_2O_3$
cc-pV5Z	-1.83	1.84	2.12	-5.64	$103.N_2O_3$

order δE_{F12} has introduced, the results still deviate from the desirable accuracy. This accuracy is reached with the addition of the δE_{INT} energy component from the interference-corrected MP2-F12 theory, as it is shown on Table 4.3. The statistics for all basis sets are greatly improved and the mean error is constantly positive, with the exception of the cc-pV5Z basis set which is negative but very close to zero (-0.20 kJ/mol).

In particular, highly accurate statistical results are obtained from the CCSD(T)-INT-F12 method with the quadruple- and quintuple-zeta basis sets. Results from the cc-pVQZ basis have a mean absolute deviation of 1.50 kJ/mol and a RMS of 1.81 kJ/mol, much lower than all the previous values from the methods neglecting the interference effects. The larger cc-pVQZ-F12 and cc-pV5Z basis drop these error below 1 kJ/mol, achieving sub-chemical accuracy in respect with the reference values. The sub-chemical accuracy was one of the main targets of the new method of this Thesis: cc-pVQZ-F12 has a MAD of 0.66 kJ/mol while the cc-pV5Z less than half a kJ/mol.

The large deviations of the double- and triple-zeta basis sets which are shown in Table 4.3 are mainly connected with the inadequate convergence behavior of the perturbative triples energy term. On the reference values, an extrapolated $\delta E_{(T)}$ energy value from quadruple- and quintuple-zeta basis was used. In order to reach the same converged accuracy, the empirical scaling of the CCSD(T*) and CCSD(T**) approaches was examined and statistics obtained from them are shown in Tables 4.4 and 4.5. Both (T*) and T(**) approaches are reducing the errors for the double- and triple-zeta basis sets, with the (T*) being always slightly more accurate than (T**) which uses the same empirical factor for both molecule and atoms. Thus, the individual scaling of molecule and atoms is more favorable. In particular, the cc-pVTZ-F12 at the CCSD(T*)-INT-F12 level of theory achieves chemical accuracy, with a RMS error of 2.11 kJ/mol. On the contrary, the atomization energies obtained from the quadruple- and quintuple-zeta basis are significantly worse with the implementation of the empirical scalings: the 0.84 kJ/mol RMS of cc-pVQZ-F12 is raised to 2.20 and 2.38 kJ/mol for (T*) and (T**), respectively. This observation is in line with the conclusions of Haunschild and Klopper [141]. In their work, they recommend the use of the CCSD(T*)(F12) method only for small and medium basis set sizes, like the cc-pVDZ-F12 and cc-pVTZ-F12 basis, while the unscaled CCSD(T)(F12) method should be preferred for accurate calculations using cc-pVQZ-F12 or larger basis sets.

Figure 4.1 shows the normal (Gaussian) distributions of the correlation component of the at-omization energies for the 106 molecules with respect to CCSD(T)(F12)/def2-QZVPP reference values (in kJ/mol). The left graph includes the distributions obtained from the cc-pVTZ-F12 basis set and the right graph from the cc-pVQZ-F12 basis set. Conventional CCSD(T), which is shown with the red lines, has for both basis sets a broad distribution of errors, significantly away from zero. The addition of the second-order correction from MP2-F12 shifts the atomiza-tion energy errors to negative values (black lines) by overestimating the energies, while they are still distributed in a wide range of errors. The Gaussians are becoming narrower only when the interference correction is being included (green lines), especially for quadruple-zeta quality basis set (ca. 4 kJ/mol range). For the cc-pVTZ-F12 basis set, there is still an important deviation from zero. On the contrary, the distribution corresponding to the cc-pVQZ-F12 basis set almost coincides with the zero value (MAD = 0.66). This sharp Gaussian has its peak at around 60, which corresponds to more than half of the total number of molecules included in the specific test set. Finally, the CCSD(T*) approach with the second-order corrections is shown with a blue line. For the quadruple-zeta quality results, the Gaussian is shifted to the left, showing also a wider base. At the same time, for the distribution of the results from the cc-pVTZ-F12 basis, even if the range is also wider (ca. 20 kJ/mol), its peak is found exactly at zero.

In all the above tables, the molecules with the maximum deviation are almost always the same. Most of them have a multi-reference character, as it is indicated by the large D1 diagnostic. [142] For example, the dinitrogen trioxide (103.N_2O_3) has a D1 value of 0.0784 and the dini-trogen tetraoxide (104.N_2O_4) 0.0692. Both D1 values have obtained from a calculation at the CCSD(T)/cc-pVQZ-F12 level of theory.

Taking into account all the above observations, some important conclusions can be made for the applicability of the interference-corrected MP2-F12 method. Firstly, the cc-pVQZ-F12 basis set provides a mean absolute deviation for the correlation energy inside the limits of sub-chemical

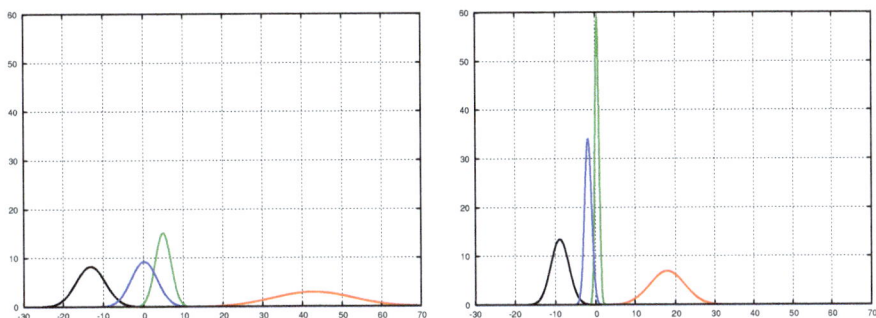

Figure 4.1: Normal distributions of atomization energies (in kJ/mol) for CCSD(T) (red), CCSD(T) plus second-order F12 energy correction (black), CCSD(T) plus second-order INT and F12 energy corrections (green) and CCSD(T*) plus second-order INT and F12 energy corrections (blue). The left graph corresponds to values obtained with the cc-pVTZ-F12 basis and the right graph from the cc-pVQZ-F12.

accuracy (0.66 kJ/mol), with respect to the basis set limit of the CCSD(T) method. In addition, a CCSD(T) calculation with a quadruple-zeta basis set is significantly less time consuming than an explicitly-correlated CCSD(T) calculation. However, the time difference between F12 and conventional coupled-cluster is larger than the time needed for the second-order corrections. This makes CCSD(T)-INT-F12 more favorable than CCSD(T)(F12). The larger cc-pV5Z basis set decreases the MAD furthermore, as it was expected, but the computational effort needed exceeds the target of this new approach. Therefore, the cc-pVQZ-F12 is proposed as the optimum basis for CCSD(T) with the second-order corrections from interference-corrected MP2-F12, since it provides the best compromise between accuracy and computational efficiency. The second conclusion has to do with the applicability of the empirical scaling of the perturbative triples. As it has already pointed out, it is advisable to be avoided for calculations with the quadruple- and quintuple-zeta basis sets, while it gains a fraction of the basis set incompleteness error of the $E_{(T)}$ for the smaller basis sets.

However, the reference correlation energy values used in the previous analysis do not represent the actual basis-set limit of the CCSD(T) model, but they are getting pretty close to it. The reason is the choice of the basis set (def2-QZVPP) which was used for the explicitly-correlated coupled-cluster calculations. This should be kept in mind in the next discussions, when a composite scheme will be presented. Nevertheless, these values consist a reasonable choice for comparison of the approximate scheme based on the interference-corrected MP2-F12, and testing the accuracy of different basis sets. The conclusions reached from the above analysis will be used in the next Sections.

As a next step, and in order to have a safer comparison, the total atomization energies of the 106 molecules computed with the INT and F12 corrections are compared with previous accurate computational results and with experimental values from the ATcT tables. For that purpose, higher-order energy components should be included, like full triples and perturbative quadruples, anharmonic zero-point vibrational energies, core/core-valence correlation, scalar relativistic and spin-orbit effects. These energy components, except from core/core-valence correlation correction, have been taken from Ref. [114] and are included in Table 4.6 under the title "Other". Table 4.6 also includes the individual components of the present composite scheme, obtained with the cc-pVQZ-F12 basis set. These components are the $E_{CCSD(T)}$ energy (which includes the HF term), the second-order corrections INT and F12, as they have been described previously and the first-

Table 4.6: Atomization energy components (in kJ/mol) of the 106-molecule test set.

Nr.[a]	Molecule		CCSD(T)	δ_{CABS}	INT[b]	F12[b]	CV	Other[c]	Total	Ref.[d]	ATcT	Error
1	CFN	Cyanogen fluoride	1255.8	0.3	-7.3	20.6	8.0	-27.9	1249.5	1250.9	1250.3 ± 1.7	-0.8
2	CFN	Isocyanogen fluoride	959.0	0.3	-6.6	19.7	6.2	-22.4	956.2	957.0	959.2 ± 2.7	-3.0
3	CF$_2$	Singlet difluoromethylene	1069.2	0.3	-4.3	14.3	2.6	-21.9	1060.2	1060.0	1059.1 ± 0.8	1.1
4	CF$_2$O	Carbonyl fluoride	1735.7	0.4	-4.9	23.4	7.1	-43.6	1718.1	1717.2	1718.4 ± 0.9	-0.3
5	CF$_4$	Tetrafluoromethane	1980.0	0.5	-5.9	25.1	5.9	-56.3	1949.3	1949.5	1947.9 ± 0.6	1.4
6	CHF	Singlet fluoromethylene	880.1	0.2	-4.5	12.1	2.3	-33.5	856.7	857.0		
7	CHFO	Formyl fluoride	1667.4	0.3	-5.8	21.2	6.5	-58.4	1631.2	1631.0	1631.4 ± 0.9	-0.2
8	CHF$_3$	Trifluoromethane	1901.9	0.4	-4.4	22.9	5.5	-75.6	1850.7	1849.2	1848.7 ± 0.9	2.0
9	CHN	Hydrogen cyanide	1289.7	0.2	-7.5	18.6	7.4	-41.7	1266.7	1268.5	1268.3 ± 0.2	-1.6
10	CHN	Hydrogen isocyanide	1227.1	0.3	-7.6	18.8	6.6	-41.2	1204.0	1205.7	1207.0 ± 0.6	-3.0
11	CHNO	Cyanic acid	1686.2	0.3	-9.6	27.0	9.4	-57.9	1655.4	1657.1	1657.2 ± 1.0	-1.8
12	CHNO	Isocyanic acid	1788.4	0.3	-8.9	27.5	9.8	-57.3	1759.8	1760.4	1761.0 ± 0.4	-1.2
13	CHNO	Formonitrile oxide	1494.8	0.3	-9.5	26.9	10.2	-50.3	1472.4	1473.9	1474.1 ± 1.2	-1.7
14	CHNO	Isofulminic acid	1437.7	0.4	-9.4	26.2	7.7	-54.7	1407.9	1409.4	1410.2 ± 1.0	-2.3
15	CH$_2$	Singlet methylene	750.0	0.1	-5.1	9.9	1.9	-42.6	714.2	715.4	714.9 ± 0.2	-0.7
16	CH$_2$F$_2$	Difluoromethane	1813.2	0.3	-6.1	20.7	5.3	-91.8	1741.6	1741.5	1741.7 ± 0.8	-0.1
17	CH$_2$N$_2$	Cyanamide	1997.5	0.4	-11.7	32.6	11.1	-89.9	1940.0	1941.5		
18	CH$_2$N$_2$	3H-Diazirine	1821.1	0.3	-11.6	31.5	8.3	-87.0	1762.6	1764.3		
19	CH$_2$N$_2$	Diazomethane	1858.0	0.4	-11.0	31.4	10.6	-80.8	1808.6	1809.9		
20	CH$_2$O	Formaldehyde	1548.4	0.2	-7.3	18.7	5.9	-70.9	1495.0	1496.6	1495.8 ± 0.2	-0.8
21	CH$_2$O	Hydroxymethylene	1332.0	0.3	-7.1	18.3	4.2	-70.8	1276.9	1277.9	1277.8 ± 1.1	-0.9
22	CH$_2$O$_2$	Dioxirane	1690.7	0.3	-8.9	25.8	5.9	-86.4	1627.4	1627.9	1629.6 ± 1.7	-2.2
23	CH$_2$O$_2$	Formic acid	2072.7	0.4	-9.3	27.4	7.8	-91.7	2007.3	2008.0	2008.4 ± 0.3	-1.1
24	CH$_2$O$_3$	Performic acid	2214.4	0.4	-10.0	33.8	8.2	-98.1	2148.7	2148.9		
25	CH$_3$F	Fluoromethane	1752.4	0.3	-7.1	18.3	5.2	-106.7	1662.4	1663.4	1665.1 ± 0.6	-2.7
26	CH$_3$N	Methanimine	1814.6	0.3	-9.6	24.0	7.2	-104.6	1731.9	1733.6	1733.5 ± 1.0	-1.6
27	CH$_3$NO	Formamide	2343.9	0.4	-11.3	33.0	9.8	-120.3	2255.5	2256.6		
28	CH$_3$NO$_2$	Methyl nitrite	2470.6	0.4	-12.4	38.2	8.3	-126.7	2378.4	2379.0		
29	CH$_3$NO$_2$	Nitromethane	2479.1	0.5	-11.8	38.9	9.9	-131.1	2385.5	2385.3		
30	CH$_4$	Methane	1743.5	0.2	-7.7	15.9	5.3	-118.6	1638.6	1640.9	1642.2 ± 0.1	-3.6
31	CH$_4$N$_2$O	Urea	3104.8	0.6	-15.8	46.9	13.2	-170.1	2979.6	2981.0		
32	CH$_4$O	Methanol	2125.9	0.3	-9.5	24.3	6.5	-136.7	2010.8	2012.4	2012.7 ± 0.2	-1.9
33	CH$_5$N	Methylamine	2408.7	0.4	-12.0	29.6	7.9	-168.4	2266.2	2268.4	2269.0 ± 0.5	-2.8
34	CO	Carbon monoxide	1072.4	0.2	-4.9	12.7	4.5	-13.9	1071.0	1072.5	1072.1 ± 0.1	-1.1
35	CO$_2$	Carbon dioxide	1607.3	0.3	-7.7	22.1	8.2	-32.8	1597.4	1599.1	1598.2 ± 0.1	-0.8
36	C$_2$F$_2$	Difluoroacetylene	1585.4	0.3	-7.4	23.1	12.6	-40.4	1573.6	1574.8	1577.0 ± 1.7	-3.4
37	C$_2$F$_4$	Tetrafluoroethylene	2432.6	0.5	-8.3	32.2	12.3	-67.4	2401.9	2402.2	2405.2 ± 1.0	-3.3

Table 4.6: *(Continued.)*

Nr.[a]	Molecule		CCSD(T)	δ_{CABS}	INT[b]	F12[b]	CV	Other[c]	Total	Ref.[d]	ATcT	Error
38	C$_2$HF	Fluoroacetylene	1640.4	0.3	-7.0	20.6	11.4	-55.9	1609.8	1610.7	1612.3 ± 1.0	-2.5
39	C$_2$HF$_3$	Trifluoroethylene	2415.0	0.5	-5.2	29.9	11.7	-84.9	2367.0	2363.5		
40	C$_2$H$_2$	Acetylene	1672.8	0.2	-7.8	18.4	10.3	-70.0	1623.9	1625.5	1626.2 ± 0.2	-2.3
41	C$_2$H$_2$F$_2$	1,1-Difluoroethylene	2423.5	0.4	-7.2	27.5	11.2	-102.0	2353.4	2352.7		
42	C$_2$H$_2$O	Ketene	2200.8	0.3	-8.8	25.7	11.9	-84.2	2145.7	2146.7	2147.3 ± 0.2	-1.6
43	C$_2$H$_2$O	Oxirene	1878.1	0.3	-10.3	26.4	10.5	-77.2	1827.8	1828.8		
44	C$_2$H$_2$O$_2$	Glyoxal	2620.4	0.3	-12.1	33.7	12.0	-99.5	2554.8	2557.1	2555.3 ± 0.6	-0.5
45	C$_2$H$_3$F	Fluoroethylene	2372.7	0.3	-8.0	25.3	10.5	-118.2	2282.6	2282.5	2278.4 ± 1.7	4.2
46	C$_2$H$_3$FO	Acetyl fluoride	2922.3	0.4	-9.2	33.4	11.9	-133.1	2825.7	2825.1		
47	C$_2$H$_3$N	Acetonitrile	2544.9	0.3	-11.4	30.7	12.8	-119.8	2457.5	2459.2		
48	C$_2$H$_3$N	Methyl isocyanide	2443.7	0.3	-11.3	30.5	11.6	-120.4	2354.4	2355.6		
49	C$_2$H$_4$	Ethylene	2334.3	0.2	-10.0	22.9	10.0	-133.5	2223.9	2226.3	2225.9 ± 0.2	-2.0
50	C$_2$H$_4$O	Acetaldehyde	2803.4	0.3	-10.8	31.0	11.3	-147.4	2687.8	2688.9	2688.9 ± 0.4	-1.1
51	C$_2$H$_4$O	Oxirane	2693.9	0.3	-12.0	31.6	11.0	-153.3	2571.5	2573.0	2573.9 ± 0.5	-2.4
52	C$_2$H$_4$O$_2$	Acetic acid	3323.0	0.5	-12.2	39.6	13.2	-166.0	3198.1	3198.5	3199.3 ± 1.5	-1.2
53	C$_2$H$_4$O$_2$	Methyl formate	3253.3	0.4	-12.0	38.4	12.5	-166.8	3126.3	3126.5	3125.2 ± 0.6	1.1
54	C$_2$H$_5$F	Fluoroethane	2988.2	0.4	-10.6	30.4	10.4	-181.6	2837.2	2838.0	2838.5 ± 1.9	-1.3
55	C$_2$H$_5$N	Aziridine	2976.4	0.4	-14.1	37.2	12.6	-186.0	2827.0	2828.7		
56	C$_2$H$_6$	Ethane	2955.0	0.3	-12.3	28.0	10.3	-198.2	2783.1	2786.0	2787.2 ± 0.2	-4.1
57	C$_2$H$_6$O	Dimethyl ether	3308.5	0.4	-13.7	35.7	11.1	-211.5	3130.5	3132.6	3132.4 ± 0.5	-1.9
58	C$_2$H$_6$O	Ethanol	3358.7	0.4	-12.8	36.3	11.6	-212.3	3181.9	3183.0	3182.8 ± 0.3	-0.9
59	C$_2$N$_2$	Cyanogen	2059.1	0.3	-12.5	32.6	15.1	-40.9	2053.7	2056.5	2055.8 ± 0.5	-2.1
60	C$_3$H$_3$N	Acrylonitrile	3147.2	0.3	-13.0	37.6	17.6	-133.2	3056.5	3057.7		
61	C$_3$H$_4$	Allene	2909.4	0.3	-12.6	29.7	15.4	-146.0	2796.2	2798.9	2800.9 ± 0.5	-4.7
62	C$_3$H$_4$	Cyclopropene	2816.7	0.3	-11.7	31.0	15.3	-148.8	2702.8	2704.3	2705.1 ± 1.0	-2.3
63	C$_3$H$_4$	Propyne	2914.9	0.3	-11.4	30.4	15.7	-146.9	2803.0	2804.1	2805.6 ± 0.5	-2.6
64	C$_3$H$_6$	Cyclopropane	3533.6	0.3	-14.1	36.1	15.7	-217.7	3353.9	3356.1	3359.7 ± 0.6	-5.8
65	C$_3$H$_6$	Propene	3565.6	0.4	-13.2	35.0	15.2	-209.5	3393.5	3395.1	3395.0 ± 0.4	-1.5
66	C$_3$H$_8$	Propane	4175.5	0.4	-15.9	39.9	15.3	-272.0	3943.2	3945.9	3944.6 ± 0.4	-1.4
67	C$_3$O$_2$	Carbon suboxide	2734.0	0.3	-11.7	35.8	18.9	-57.9	2719.4	2721.3		
68	C$_4$H$_4$	Butatriene	3491.4	0.3	-14.2	36.8	20.5	-156.9	3377.9	3380.1		
69	C$_4$H$_4$	Cyclobutadiene	3386.6	0.4	-15.1	38.8	18.9	-159.8	3269.8	3272.3		
70	C$_4$H$_4$	Tetrahedran	3272.1	0.4	-15.3	40.0	21.6	-160.3	3158.5	3160.4		
71	C$_4$H$_4$	Vinylacetylene	3526.4	0.3	-13.4	37.3	20.6	-160.8	3410.4	3411.6		
72	C$_4$N$_2$	Dicyanoacetylene	3265.3	0.3	-15.9	47.0	26.0	-68.3	3254.4	3256.1		
73	FH	Hydrogen fluoride	587.1	0.2	-2.9	6.7	1.0	-26.8	565.3	565.9	566.0 ± 0.0	-0.7
74	FHO	Hypofluorous acid	652.2	0.2	-4.9	12.8	1.2	-37.4	624.1	624.3	624.0 ± 0.4	0.1

Table 4.6: *(Continued.)*

Nr.[a]	Molecule	CCSD(T)	δ_{CABS}	INT[b]	F12[b]	CV	Other[c]	Total	Ref.[d]	ATcT	Error
75	Fluoroperoxide	860.4	0.3	-6.6	19.4	1.6	-47.2	827.9	827.4		
76	Monofluoroamine	1057.1	0.3	-7.5	19.2	2.5	-73.3	998.3	999.0		
77	Fluorohydrazine	1686.3	0.4	-11.2	32.7	5.5	-119.4	1594.3	1594.1		
78	Nitrosyl fluoride	880.8	0.3	-6.4	20.1	2.3	-17.3	879.8	878.7		
79	Difluorine	156.4	0.1	-2.5	6.0	0.1	-6.0	154.1	153.8	154.6 ± 0.2	-0.5
80	Difluorodiazene (cis)	1029.1	0.3	-8.2	27.3	3.5	-31.2	1020.8	1019.7		
81	Difluorodiazene (trans)	1023.3	0.3	-8.4	27.2	3.6	-31.1	1014.9	1013.5		
82	Difluorine monoxide	379.6	0.2	-3.9	12.8	0.4	-13.7	375.4	373.5	373.3 ± 0.7	2.1
83	Perfluoroperoxide	616.2	0.3	-5.4	19.3	0.7	-15.7	615.4	612.3	609.7 ± 0.8	5.7
84	Trifluoroamine	844.3	0.3	-4.6	21.5	1.2	-30.5	832.2	828.9		
85	Nitrosylhydride	844.1	0.3	-7.5	18.7	2.7	-35.3	823.0	824.3	823.6 ± 0.1	-0.6
86	Nitrous acid (cis)	1281.2	0.4	-9.4	26.6	3.7	-52.0	1250.5	1251.7	1251.5 ± 0.4	-1.0
87	Nitrous acid (trans)	1282.6	0.4	-9.1	26.5	3.8	-51.6	1252.6	1253.0	1253.3 ± 0.1	-0.7
88	Nitrous acid, H-NO$_2$	1249.3	0.4	-8.9	27.2	4.9	-57.1	1215.8	1216.4		
89	Nitric acid	1591.7	0.5	-9.4	34.7	6.1	-70.9	1552.7	1552.0	1551.6 ± 0.2	1.1
90	Hydrogen azide	1353.4	0.5	-11.5	32.7	7.8	-53.5	1329.4	1330.8	1329.7 ± 0.6	-0.3
91	Diazene (cis)	1195.9	0.4	-10.0	24.8	4.1	-71.5	1143.7	1145.4	1143.5 ± 0.9	0.2
92	Diazene (trans)	1218.4	0.4	-10.2	24.8	4.2	-73.1	1164.5	1166.1	1165.8 ± 0.7	-1.3
93	Diazene (iso)	1114.0	0.3	-9.7	25.2	4.7	-69.3	1065.2	1066.1	1065.1 ± 0.9	0.1
94	Nitrosamide	1586.8	0.4	-11.7	32.8	6.2	-84.2	1530.3	1531.4		
95	Water	964.3	0.3	-5.7	12.7	1.9	-57.1	916.4	917.6	917.8 ± 0.1	-1.4
96	Hydrogen peroxide	1109.6	0.3	-8.2	19.4	2.4	-70.2	1053.3	1055.0	1055.2 ± 0.1	-1.9
97	Ammonia	1231.7	0.3	-8.0	17.9	3.1	-90.0	1155.0	1156.9	1157.3 ± 0.1	-2.3
98	Ammonia oxide	1380.4	0.4	-9.7	25.9	3.9	-108.8	1292.1	1290.5		
99	Hydroxylamine	1483.8	0.4	-9.8	25.4	3.9	-106.4	1397.3	1398.7	1398.7 ± 0.5	-1.4
100	Hydrazine	1806.9	0.5	-12.9	31.2	5.7	-139.9	1691.5	1694.0	1695.6 ± 0.2	-4.1
101	Dinitrogen	936.4	0.2	-7.5	18.9	4.3	-12.7	939.6	940.8	941.1 ± 0.1	-1.5
102	Nitrous oxide	1104.1	0.4	-9.2	26.9	6.4	-27.3	1101.3	1102.8	1102.0 ± 0.1	-0.7
103	Dinitrogen trioxide	1598.0	0.5	-10.7	40.4	6.6	-41.3	1593.5	1592.4	1591.1 ± 0.2	2.4
104	Dinitrogen tetraoxide	1923.9	0.6	-15.9	48.7	9.1	-60.5	1905.9	1908.0	1908.5 ± 0.2	-2.6
105	Ozone	590.1	0.3	-6.5	18.5	1.4	-8.2	595.6	594.9	596.1 ± 0.1	-0.5
106	Dihydrogen	456.2	0.1	-2.4	4.0	0.0	-26.1	431.8	432.7	432.1 ± 0.0	-0.3

[a] Same number and same molecule as in Ref. [33] except for dihydrogen.
[b] Second-order corrections from INT-MP2-F12 (approximation A).
[c] Corrections to the electronic energy taken from Ref. [114]. See text for further explanation.
[d] Total energy from Ref. [114].

Table 4.7: Basis-set convergence of the core/core-valence contribution (in kJ/mol) as obtained at the CCSD(T) level. The ROHF reference was used for the atoms.

Nr.[a]	Molecule		cc-pCVQZ[b]	cc-pCV5Z[b]	cc-pCVQZ-F12	cc-pCV(Q5)Z[b]
1	CFN	Cyanogen fluoride	6.92	7.26	7.21	7.61
9	CHN	Hydrogen cyanide	6.59	6.95	6.82	7.32
10	CHN	Hydrogen isocyanide	5.78	6.06	5.98	6.35
15	CH_2	Singlet methylene	1.59	1.65	1.65	1.71
20	CH_2O	Formaldehyde	5.21	5.43	5.36	5.65
30	CH_4	Methane	4.99	5.19	5.06	5.40
34	CO	Carbon monoxide	3.76	3.96	3.90	4.17
35	CO_2	Carbon dioxide	7.00	7.33	7.29	7.68
40	C_2H_2	Acetylene	9.60	10.10	9.86	10.62
73	FH	Hydrogen fluoride	0.77	0.77	0.75	0.77
74	FHO	Hypofluorous acid	0.64	0.62	0.67	0.60
79	F_2	Difluorine	-0.29	-0.32	-0.28	-0.35
92	H_2N_2	Diazene (trans)	3.34	3.44	3.53	3.56
95	H_2O	Water	1.61	1.63	1.61	1.66
97	NH_3	Ammonia	2.71	2.80	2.78	2.88
101	N_2	Dinitrogen	3.38	3.55	3.57	3.74
102	N_2O	Nitrous oxide	5.03	5.21	5.37	5.41
105	O_3	Ozone	0.24	0.17	0.38	0.09

[a] Same number and same molecule as in Ref. [33] except for dihydrogen.
[b] Data taken from Ref. [114].

order correction to the HF energy obtained from the explicitly-correlated MP2-F12 method (CABS singles), which corresponds to the column under the name "δ_{CABS}". CABS singles is an extra term based on the single-excitations into the virtual orbitals in a formally complete basis and contributes to the basis set incompleteness error of the zeroth-order wave function and thus, it is a correction to the HF energy. [109]

Additionally, the core/core-valence contributions to the atomization energies are also present in Table 4.6 ("CV" column). This correction was obtained from the difference between all-electron and frozen-core CCSD(T) calculations with the cc-pCVQZ-F12 basis (Eq. (4.2)). The cc-pCVQZ-F12 basis has been optimized specifically for accurately describing core-core and core-valence correlation effects with explicitly correlated F12 methods and they have been developed by augmenting the cc-pVQZ-F12 and aug-cc-pVQZ families of basis sets with additional functions whose exponents were optimized based on the difference between all-electron and valence-electron correlation energies. [135] In contrast with the rest, higher-order corrections which were obtained from Ref. [114], this correction was recomputed due to the higher accuracy that the cc-pCVXZ-F12 basis sets offer.

In order to verify the applicability of the specific basis set for the CV correction in the current composite scheme, a few (18) representative cases were compared. Table 4.7 includes the E_{CV} energy term for these 18 molecules. In this comparison, the core/core-valence correction was calculated with the cc-pCVQZ, cc-pCV5Z and cc-pCVQZ-F12 basis sets. The basis set limit of this term has obtained from a two-point X^{-3} extrapolation based on the cc-pCVQZ and cc-pCV5Z results and it is labeled as cc-pCV(Q5)Z. For the atoms, a restricted open-shell Hartree-Fock (ROHF) reference was used. The core/core-valence correction evaluated from the cc-pCVQZ-F12 basis is for almost all the cases in the same magnitude with the quintuple-zeta quality results and

Table 4.8: Statistics of the deviations (in kJ/mol) of the computed values obtained from different schemes with respect to the high accurate computational data of Ref. [114]. In all these models, the cc-pVQZ-F12 basis set was used.

Method	Approx[a]	Mean Error	MAD	RMS	Max Error	Molecule
UCCSD(T)	A	0.84	1.31	1.51	3.4	$39.C_2HF_3$
UCCSD(T*)	A	-1.44	1.71	2.33	7.4	$83.F_2O_2$
UCCSD(T)	B	1.97	2.13	2.37	4.3	$104.N_2O_4$
UCCSD(T*)	B	-0.17	1.23	1.66	6.2	$83.F_2O_2$
UCCSD(T)	AB	1.29	1.60	1.81	3.4	$59.C_2N_2$
UCCSD(T*)	AB	-0.99	1.43	2.02	7.0	$83.F_2O_2$
ROHF-CCSD(T)	A	1.47	1.66	1.84	3.9	$104.N_2O_4$
ROHF-CCSD(T*)	A	-0.77	1.24	1.74	6.0	$83.F_2O_2$
ROHF-CCSD(T)	B	2.55	2.61	2.83	6.0	$104.N_2O_4$
ROHF-CCSD(T*)	B	0.43	1.19	1.49	4.9	$83.F_2O_2$
ROHF-CCSD(T)	AB	1.90	2.02	2.22	4.7	$104.N_2O_4$
ROHF-CCSD(T*)	AB	-0.34	1.11	1.54	5.6	$83.F_2O_2$

[a] Approximation used for the INT-MP2-F12. See text for details.

always smaller (in absolute values) from the extrapolated energies. Thus, this specific basis set provides a more accurate CV term to the composite scheme than the cc-pCVQZ values. However, its size is comparable to the quintuple-zeta cc-pCV5Z basis, making the specific correction more computational demanding.

If the cc-pVQZ-F12 basis set which was used for the interference-corrected MP2-F12 was also applied in the CV correction, then the obtained energy contributions would be significantly larger than the basis-set limit. The reason for that behavior is the lack of the additional polarization functions, which are important for the core-core and core-valence region of correlation energy. This behavior is a form of basis-set superposition error. For example, the CV correction for the cyanogen fluoride (1.CFN), as it is obtained from the cc-pVQZ-F12 basis, is 8.51 kJ/mol, which is about 0.9 kJ/mol larger than the basis-set limit.

Finally, in Table 4.6, the column under the name "Total" includes the total atomization energies after the summation of the individual terms. In the column "Ref" the computational atomization energies from Ref. [114] are shown. The last two columns include the experimental available atomization energies from the ATcT tables and the relative error of the new composite scheme proposed in this Thesis.

A comparison between the highly accurate composite scheme of Klopper et al. [114] and the different approximations and empirical corrections which can be added to CCSD(T) energies with the second-order corrections from the interference-corrected MP2-F12 leads to important conclusions for the accuracy which can be obtained from the "one-basis" approach of the CCSD(T)-INT-F12 method. These conclusions are based on which of the proposed alternative models can achieve higher accuracy. The target is to formulate a computational cheaper protocol for the calculation of atomization energies which can still provide accurate results. Apart from the (T*), the empirical scaling of the perturbative triples, which has been already discussed in Section 4.1.3, the different approximations of the MP2-F12 theory are also examined. These are either the approximation A or B, or a 60% - 40% weighted sum of both, respectively, which will be mentioned further on as "AB". These averaged energies have been proposed by Samson and Klopper [117], based on the convergence behavior of MP2-F12/A and MP2-F12/B methods. The advantages of this approach has been discussed in Section 3.3.4. One additional consideration which was taken into account for the post-HF calculations of the atoms is the reference wave function. These can be the unrestricted or the restricted open-shell wave function and will be mentioned as UCCSD(T) and ROHF-CCSD(T), respectively. The variety of the alternative combinations of these models with their corresponding statistics are shown in Table 4.8, as they have been computed with the cc-pVQZ-F12 basis set.

The most accurate scheme is the UCCSD(T)/A which uses the interference-corrected MP2-F12 with approximation A and an unrestricted HF reference for the atoms. This composite scheme includes no empirical scaling for the perturbative triples nor an averaged sum of approximations A and B and it yields a mean absolute deviation of 1.31 kJ/mol and an RMS value of 1.51 kJ/mol. If the (T*) scaling is used, then the corresponding deviations are significantly larger. This observation agrees with the previous conclusions, when only the correlation energies of the CCSD(T)-INT-F12 and CCSD(T)(F12) methods were compared. The (T**) which was mentioned in Section 4.1.3 gives similar results as the schemes which includes individual (T*) scalings for the molecule and the atoms. Therefore, the (T**) scaling is not mentioned in this comparison.

Approximation B also yields larger deviations than A but these errors are compensated by the addition of the (T*) correction. The mean error of the unscaled model drops from about 2 to almost 0 kJ/mol, when the empirical scaling is added. The UCCSD(T*)-INT-F12/B scheme has an RMS error of 1.66 kJ/mol, in respect with the highly accurate results of Ref. [114]. The weighted approximations A and B (AB), with or without the (T*) provide results between the values of pure approximation A and B. Even if for some specific cases approximation B and AB may give more accurate results from approximation A, statistically the smallest deviations are obtained from the UCCSD(T)-INT-MP2/A method.

Almost similar behavior was observed for the approximations A, B and AB when the restricted open-shell Hartree-Fock was used as reference wave function for the atoms. The main difference from the UCCSD(T) models is observed on the models which include the empirical scaling of the perturbative triples. All these three models result smaller errors than the unscaled variants. In fact, errors from the ROHF-CCSD(T*) with interference-corrected MP2-F12 which uses approximation B or AB are lower than the corresponding values from the ROHF-CCSD(T)-INT-F12/A model and comparable with the unscaled UCCSD(T)-INT-F12/A deviations: ROHF-CCSD(T*)-INT-F12/B has an RMS error of 1.49 kJ/mol and ROHF-CCSD(T*)-INT-F12/AB of 1.54 kJ/mol. For the latter model, the MAD of 1.11 kJ/mol is the smallest from all the approaches taken into account in Table 4.8.

As far as no significant improvement is gained from the combinations of the (T*) and/or AB empirical scalings when they are used for correcting the UCCSD(T) or ROHF-CCSD(T) atomization energies, the UCCSD(T)-INT-F12/A model is suggested as the most reasonable choice for a model chemistry based on the interference-corrected MP2-F12. The three models with the lowest RMS errors which were discussed before, UCCSD(T)-INT-MP2/A, ROHF-CCSD(T*)-INT-F12/B and ROHF-CCSD(T*)-INT-F12/AB, achieve almost similar accuracy. However, even if their RMS values are around 1.5 kJ/mol, they differ in the size of their maximum errors. The two restricted open-shell models have a maximum error for perfluoroperoxide (83.F_2O_2). When the INT-MP2-F12/B is used, the difference from the reference computational value of Klopper *et al.* [114] is 4.9 kJ/mol, while the application of the averaged AB approach increases this error to 5.6 kJ/mol. On the contrary, the UCCSD(T)-INT-MP2/A model has a deviation for perfluoroperoxide of 3.1

Table 4.9: Statistics of the deviations (in kJ/mol) of the computed values with respect to the ATcT reference data.

F12 contribution	Mean Error	MAD	RMS	Max Error	Molecule
CCSD(F12)[a]	-0.1	1.3	1.7	3.4	83.F_2O_2
CCSD(F12/fixed)[a]	-0.5	1.2	1.5	3.0	64.C_3H_6
MP2-F12($f_{int} = 0.0$)[b]	-21.7	21.7	23.1	46.3	104.N_2O_4
MP2-F12($f_{int} = 0.78$)[b]	-0.12	0.90	1.22	4.1	45.C_2H_3F
MP2-F12($f_{int} = 1.0$)[b]	5.98	5.98	6.52	14.1	104.N_2O_4
INT-MP2-F12	1.16	1.75	2.15	5.7	64.C_3H_6
INT-MP2-F12 (with (T*))	-0.92	1.71	2.50	10.0	83.F_2O_2

[a] Data taken from Ref. [132].
[b] Data taken from Ref. [114].

kJ/mol, and having its maximum error for the trifluoroethylene (39.C_2HF_3). This maximum error (3.4 kJ/mol) is significantly smaller than the corresponding errors of the two other models (more than 1.5 kJ/mol). Thus, the UCCSD(T)-INT-MP2/A yields consistently smaller deviations. In addition, the empirical scaling of the perturbative triples and the 60%-40% average of approximations A and B are avoided. This leads to the conclusion that the composite scheme based on the UCCSD(T) with second-order corrections from the interference-corrected MP2-F12 with approximation A yields the most accurate atomization energies from the models discussed above, in respect to the values from Ref. [114]. Additionally, it does not fall back on empirical factors while at the same time avoids the computational demanding CCSD(T)(F12) calculations as performed in Ref. [132].

As a final step, a discussion of the results of the previous accurate computational studies [114, 132] and the models which include the CCSD(T)-INT-F12 method should be given. This comparison has been done with respect to the experimental values of the ATcT reference data and it shows the potential applicability of the new composite scheme's protocol as a computational cheaper but still accurate method. The statistics of these methods are shown in Table 4.9. The first two lines correspond to the composite schemes which calculate the basis-set limit of the CCSD(T) method from the explicitly-correlated coupled-cluster theory. [132] The specific schemes provides statistically very accurate atomization energies for the 106-molecule test set. This is clear from the RMS deviations of 1.7 and 1.5 kJ/mol for the explicitly-correlated variant which optimizes the F12 amplitudes and the fixed-amplitudes approach, respectively. Different accuracy is achieved when the basis-set limit of the coupled-cluster method is approached from second-order corrections. Plain CCSD(T) without any second-order corrections ($f_{int} = 0.0$) has a very large mean absolute deviation of 21.7 kJ/mol. An unscaled contribution of the F12 term, i.e. $f_{int} = 1.0$, performs better than the pure CCSD(T), but still worse than the explicitly-correlated CCSD(T) results, with a 6.52 kJ/mol RMS value, while an empirically scaled F12 contribution ($f_{int} = 0.78$), with a RMS of 1.22 kJ/mol is significantly more accurate. The MAD of 0.90 kJ/mol is a clear indication of the accuracy which can be achieved from the second-order corrections to the CCSD(T) energy.

Table 4.9 includes also the statistics with respect to the experimental values of the ATcT reference data of the models with the UCCSD(T) and interference-corrected MP2-F12 energy components, with and without the empirical scaling of the perturbative triples. These values have been calculated with the cc-pVQZ-F12 basis set. As it was expected from the comparison between these models and the accurate computational results of Klopper et al. [114], account of the (T*) correction increases the error for this specific model. However, all models which were discussed before, i.e. with unrestricted or restricted open-shell reference wave function, with approximation A, B or averaged AB and with or without the (T*) correction, yield an RMS error less than 1 kcal/mol. Among them, the most accurate models are the UCCSD(T)-INT-F12/A, the ROHF-CCSD(T*)-INT-F12/B and the ROHF-CCSD(T*)-INT-F12/AB. All these three models yield atomization energies with very small deviations from the ATcT values and their RMS errors are close to 2.10 kJ/mol. However, like it has been mentioned in the previous discussion, the UCCSD(T)-INT-F12/A model is more favorable than the rest because it does not include any empirical scalings and it has again the lowest maximum deviation (5.7 kJ/mol for cyclopropane) of all models.

4.1.5 Additivity scheme

The 106-molecule test set was also used for comparison between the different basis set pairs ("dual level") of the additivity scheme. The main idea and the details of this scheme have been described in Section 3.4. In this section, its applicability will be discussed, with or without the empirical (T*) correction, and the results will be compared with the CCSD(T)(F12)/def2-QZVPP correlation energy, the high-accurate atomization energies of Ref. [114] and with the ATcT experimental values.

The different combinations of basis sets which were taken into account are the cc-pVXZ basis ($X = D, T, Q, 5$) of Dunning [19] as "small" basis (SB) and the cc-pVXZ-F12 basis ($X = D, T, Q$) of Peterson and co-workers [103] as "large" basis (LB). It should be noted that the notation SB and LB is somehow naive, as far as for combinations like the cc-pV5Z/cc-pVDZ-F12 as SB/LB, respectively, which have also been examined, the "SB" quintuple-zeta quality basis is much larger in number of basis functions than the "LB" double-zeta cc-pVDZ-F12 basis set. However, this notation is kept through the whole discussion, in order to separate the contributions obtained from the different basis sets, as shown on Eq. (3.22). The SB is used for the CCSD(T) energy and the calculation of the interference factor which will scale the second-order contributions, as they are calculated from the difference between the $\delta E_{\text{MP2-F12/LB}} - \delta E_{\text{MP2/SB}}$ energies. This notation also offers a consistency through the analysis of the results.

Apart from the different basis set combinations, four different models are compared. The first two do not take into account any interference effects and correspond to the method of Eq. (3.21) where the F^{ij} factor is set to one. The difference between these two models is the use or not of the empirical (T*) scaling. The other two models include the calculated interference factor explicitly for every electron pair, with or without the (T*) contribution. In addition, the two different approximations A and B of the explicitly-correlated MP2 are also discussed.

Firstly, the correlation energy component of the atomization energies of the 106 molecules will be analyzed. These results are compared with the CCSD(T)(F12)/def2-QZVPP correlation energy, as it has been done in the beginning of the previous section for the one-basis scheme of the CCSD(T)-INT-F12 method. All the statistical results between the different basis set combinations, methods and F12 approximations are shown in Tables D.1 to D.4 in Appendix D.

All results which include the cc-pVDZ basis set as SB are shown in Table D.1. For these methods, the RMS error is always above 10 kJ/mol. This means that the specific basis set combinations are not recommended for highly accurate thermochemistry results. The main reason is that the CCSD(T) correlation energy calculated from a double-zeta quality basis is quite poor and even with all the additional correcting terms, it can not provide reliable atomization energies. However, some interesting comments can be made by having a closer look at Table D.1. Firstly, the empirical scaling of the (T) does not always corrects the correlation energy obtained from the

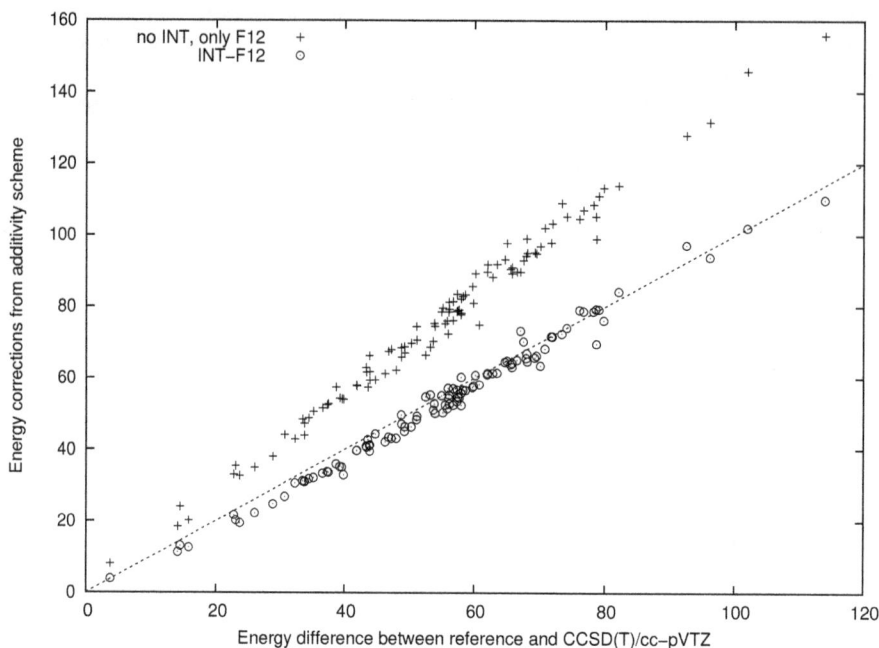

Figure 4.2: Comparison between the errors from the UCCSD(T)+F12/A (+) and UCCSD(T)-INT-F12/A (o) additivity schemes. Both schemes correspond to the cc-pVTZ/cc-pVDZ-F12 basis set combination. All values in kJ/mol

additivity scheme. It does when the interference-corrected MP2-F12 is used but it deteriorates the statistical results when $F^{ij} = 1$. This trend holds for all the three different LB sets. Secondly, approximation B performs better for the model without the INT energy component while for the UCCSD(T)-INT-F12 and UCCSD(T*)-INT-F12 models, it yields slightly worse results. This remark agrees also with the atomization energies obtained from the one-basis approach where the approximation A was superior than B, in terms of energy deviations. Finally, the best results have been calculated from the cc-pVDZ/cc-pVDZ-F12 combination of basis sets when the second-order interference and F12 terms are included (approximation A) in combination with the (T*) scaling. This observation is also in line with the conclusions of Section 4.1.4, where CCSD(T)-INT-F12 with the double-zeta quality basis sets was providing more accurate results when the contribution of the empirical (T*) was added.

The accuracy of the additivity scheme is significantly improved when a triple-zeta basis set is used as small basis (Table D.2). Corrections to the CCSD(T)/cc-pVTZ correlation energy are performing fairly well, especially when the INT and F12 terms are calculated with the approximation A. In particular, two models, the UCCSD(T)-INT-F12/A with the cc-pVTZ/cc-pVDZ-F12 basis set combination and the UCCSD(T*)-INT-F12/A with the cc-pVTZ/cc-pVTZ-F12 basis set combination, result in an RMS error close to 3 kJ/mol. Between these two models, the first is preferred because it does not fall back on the empirical (T*) scaling, it uses a "smaller" basis set as LB and the maximum error of it (9.1 kJ/mol for carbon suboxide) is significantly smaller than the corresponding maximum error of the second model (12.4 kJ/mol for dinitrogen trioxide).

From Table D.2 and for approximation A, the importance of the interference effects is also clear.

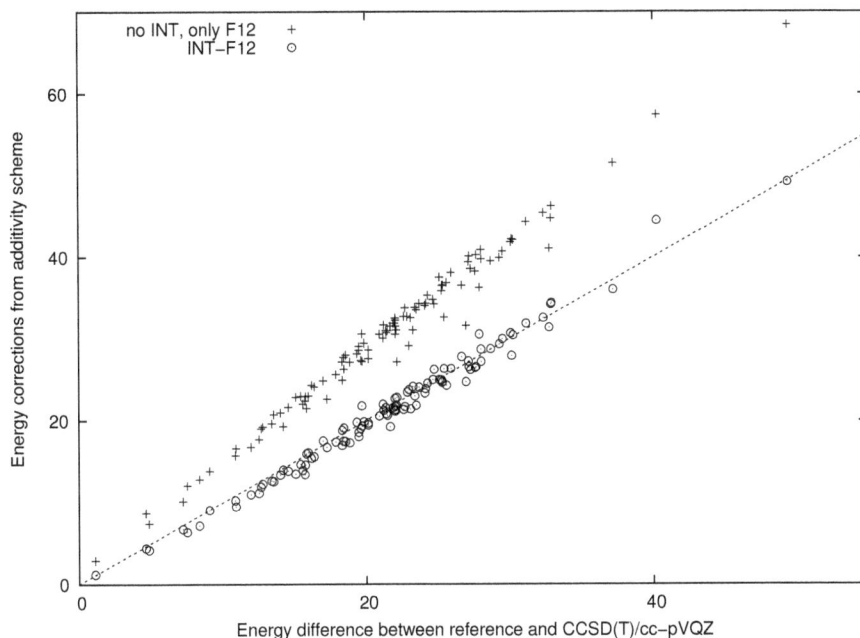

Figure 4.3: Comparison between the errors from the UCCSD(T*)+F12/B (+) and UCCSD(T*)-INT-F12/B (○) additivity schemes. Both schemes correspond to the cc-pVQZ/cc-pVTZ-F12 basis set combination. All values in kJ/mol

These effects contribute to the decrease of errors between the additivity scheme and the reference values. In Figure 4.2 the second-order corrections of the additivity scheme for the cc-pVTZ/cc-pVDZ-F12 basis set combination are plotted against the difference between the reference values and the CCSD(T)/cc-pVTZ energies. This difference corresponds to the basis set incompleteness error which the INT-MP2-F12 recovers. The crosses (+) correspond to the UCCSD(T)+F12/A model, in which no interference effects are present ($F^{ij} = 1$). All points divert significantly and are almost lying on a straight line which is placed above the "optimum" diagonal. This has as a result a very high mean absolute deviation of 21.08 kJ/mol for this specific model. Addition of the interference correction shifts all these values almost on the diagonal, as shown as circles (○) in Figure 4.2. This means that the UCCSD(T)-INT-F12/A scheme recovers the missing fraction of the correlation energy and corresponding errors drop notably: the MAD of this specific scheme with the cc-pVTZ/cc-pVDZ-F12 basis set combination is reduced to 2.59 kJ/mol.

Another interesting remark is that the MAD and RMS errors of the UCCSD(T*)-INT-F12/A models which use the cc-pVTZ basis as SB are not reduced in analogy with the increase of the size of the LB. On the contrary, they yield the minimum error for the cc-pVTZ/cc-pVTZ-F12 basis set combination. However, the maximum error follows the expected decrease from the cc-pVDZ-F12 to the larger cc-pVQZ-F12 basis. Additionally, for the unscaled UCCSD(T)+F12 and UCCSD(T)-INT-F12 models (approximation A or B) the same statistical deviations are increased when the size of the LB is also increased. The above remarks are clear indications that the statistics obtained from the additivity schemes with the cc-pVTZ as small basis, especially the ones from the UCCSD(T)-INT-F12/A model with the cc-pVTZ/cc-pVDZ-F12 basis set combination, are based on cancellation of errors between the individual terms. Nevertheless, the specific model can be proposed for the calculation of the correlation energy of a computational cheap model chemistry.

Higher accuracy is achieved when the cc-pVQZ basis set is used as small basis. The corresponding results are shown on Table D.3. This accuracy is mainly due to the first and dominating term (i.e. corresponds to a higher percentage of the correlation energy than the corrections) of the right-hand side of Eq. (3.22). With the quadruple-zeta quality basis, this term is calculated with higher precision and thus, the contribution of further corrections is smaller.

For the models which do not include any interference effects but use the approximation A for the F12 component, the RMS errors are dropped when the LB is increased. The opposite behavior is observed when the approximation B is used: from a 4 kJ/mol RMS value for the cc-pVQZ/cc-pVDZ-F12 combination and for the unscaled (T) contributions the error increases to 7 kJ/mol for the cc-pVQZ/cc-pVQZ-F12 combination of basis sets. On the contrary, and as it is expected, the MAD and RMS values decrease for the UCCSD(T)-INT-F12 models when the size of the LB is increased, with or without the (T*) correction. However, this trend is not followed for the UCCSD(T)-INT-F12/A model: the 1.23 kJ/mol RMS value of the cc-pVQZ/cc-pVTZ-F12 combination is significantly lower than the corresponding one of the cc-pVQZ/cc-pVQZ-F12 combination. Even if this observation is due to error cancellation, the maximum deviation of this specific model for the trifluoroethylene molecule (4.2 kJ/mol) is surprising small, making this specific model also a candidate for model chemistries. Similar accuracy is achieved from the UCCSD(T*)-INT-F12/B models with cc-pVQZ/cc-pVTZ-F12 (RMS = 1.06 kJ/mol, maximum error 4.3 kJ/mol) and cc-pVQZ/cc-pVQZ-F12 (RMS = 1.06 kJ/mol, maximum error 5.0 kJ/mol) basis set combinations, but both models have the drawback that they include the empirical (T*) term.

Nevertheless, the low RMS obtained from these models is due to the addition of the interference effects. Figure 4.3 includes the second-order corrections with (○) and without (+) the E_{INT} energy component for the models which include the (T*) correction, the approximation B for the F12 term and the cc-pVQZ/cc-pVTZ-F12 basis set combination against the basis set incompleteness error of the CCSD(T)/cc-pVQZ level of theory. The second-order corrections which lack the interference effects (+) are lying in an almost a straight line, significantly above the diagonal. On the contrary, the INT terms are correcting this behavior by regaining the energy difference between the reference correlation energies and the CCSD(T)/cc-pVQZ energies. This behavior is reflected on the MAD of these two models; from almost 9 kJ/mol, the mean absolute deviation drops below 1 kJ/mol (0.84 kJ/mol).

Smaller errors and deviations from the reference correlation energies are obtained when the larger cc-pV5Z basis is used as SB in the additivity schemes (Table D.4). The interference effects on the schemes which include the cc-pVDZ-F12 basis seem to lead to larger deviations from the reference values. On the contrary, the schemes with the cc-pVTZ-F12 and cc-pVQZ-F12 basis as LB are significantly more accurate when the INT term is included. In specific cases, the RMS error is smaller than 1 kJ/mol and the maximum deviations does not exceed 3 kJ/mol. In particular, the UCCSD(T)-INT-F12/A model with the larger basis set combination included in this study (cc-pV5Z/cc-pVQZ-F12) yields a mean absolute error of 0.60 kJ/mol and an RMS value of 0.85 kJ/mol, when the maximum error is only 2.0 kJ/mol for the trifluoroethylene molecule (C_2HF_3). Finally, and as it was expected, all models which use as SB the quintuple-zeta quality basis have larger deviations when the (T*) scaling is included.

A choice between the different additivity schemes discussed in the previous paragraphs and showed in Tables D.1 to D.4 in Appendix D should be done in order to proceed with suggestions of composite schemes for calculation of atomization energies, like it has been done for the one-basis approach (Section 4.1.4). The target of the composite scheme is to include a relative cheap computational method which will approximate as accurate as it can the CBS correlation energy of the CCSD(T) method. Thus, the calculation of the CCSD(T) energy, which corresponds to the first term of the right-hand side of Eq. (3.22), should be kept as cheap as possible, in terms of computational time. Thus, the use of the quintuple-zeta quality basis should be avoided. The other alternatives for SB are the cc-pVTZ or cc-pVQZ basis sets which can calculate also with high accuracy (RMS error less than 1 kcal/mol) the correlation component of the atomization energies. In particular, the UCCSD(T)-INT-F12/A and UCCSD(T*)-INT-F12/B schemes with the cc-pVQZ/cc-pVTZ-F12 combination of SB/LB sets, respectively, will be taken into account, due to their low deviations from the reference values. Furthermore, a more "compact" additivity scheme which will avoid the coupled-cluster calculation with the quadruple-zeta quality basis will be examined. This will include the cc-pVTZ basis as SB and they will be based on the UCCSD(T)-INT-F12/A scheme with the cc-pVTZ/cc-pVDZ-F12 basis or on the UCCSD(T*)-INT-F12/A with the cc-pVTZ/cc-pVTZ-F12 basis set combination. The choice has been made due to the good performance of both schemes, as they are shown on the statistical results of Table D.2. It should also be mentioned that the averaged contribution to the second-order corrections which was examined in the previous section (i.e. 60% from approximation A, 40% from approximation B) was not taken into account in the additivity schemes. The reason is that there was not observed important correction to the correlation energy for the one-basis approach and it is believed that it will not contribute either in the two-basis models.

The composite methods based on the additivity scheme are similar with the one described in the previous section. The HF energy was obtained from the MP2-F12 calculation which corresponds to the large basis set. Two are the reasons for this choice. The first is that for the most of the basis set combinations, the LB has more basis functions than the SB and thus, the HF energy is closer to the basis set limit. This holds for the schemes where the LB has the same or larger cardinal number from SB. When both basis share the same cardinal number, like for example triple-zeta, then the cc-pVTZ-F12 (LB) basis is larger than the cc-pVTZ, which lacks the extra polarization functions. In that case, the HF contribution from the cc-pVTZ-F12 basis to the atomization energy is larger than the one from the cc-pVTZ basis. The second reason is that the explicitly-correlated MP2 calculation provides also the CABS singles term which corrects the basis set incompleteness of the HF energy. In addition, the same CV correction was added like in the one-basis composite scheme, as it has been calculated with the cc-pCVQZ-F12 basis set. This basis set provides a value close to the basis set limit of the term, as it has been shown in Table 4.7. Finally, the extra corrections (under the name "Other") which were also added to the previous composite scheme and have been described above are included.

The upper part of Table 4.10 presents the statistics of the four models which provided the most accurate results for the correlation part of the atomization energies of the 106 molecules. These statistics are with respect to the experimental ATcT reference data. The mean absolute deviations of the two models which use the cc-pVTZ basis set (SB) for the CCSD(T) energy exceed 1 kcal/mol. Their RMS errors are 5.24 kJ/mol, for cc-pVDZ-F12 as LB, and 4.60 kJ/mol, for cc-pVTZ-F12. These errors are significantly lower when a basis of quadruple-zeta quality is used as SB. The two cc-pVQZ/cc-pVTZ-F12 models (or "QT" in a shorthand notation) under investigation have an RMS value of 1.90 kJ/mol, when the approximation A is used for the INT-MP2-F12 calculation without the empirical (T*) correction, and 2.64 kJ/mol, when the the approximation B with the (T*) correction is added. The lower part of Table 4.10 includes some extra models

Table 4.10: Statistics of the deviations (in kJ/mol) of the computed values from the additivity schemes with respect to the ATcT reference data.

Basis[a]	Model	Mean Error	MAD	RMS	Max Error	Molecule
TD	UCCSD(T) -INT-F12/A	-4.21	4.66	5.24	11.0	61.C_3H_4
TT	UCCSD(T*)-INT-F12/A	-1.51	3.62	4.60	13.9	103.N_2O_3
QT	UCCSD(T) -INT-F12/A	-0.55	1.47	1.90	5.3	45.C_2H_3F
QT	UCCSD(T*)-INT-F12/B	-1.27	2.12	2.64	7.3	83.F_2O_2
QQ	UCCSD(T*)-INT-F12/A	0.44	1.64	2.38	8.7	103.N_2O_3
QQ	UCCSD(T*)-INT-F12/B	-0.71	1.76	2.33	7.2	83.F_2O_2
5T	UCCSD(T) -INT-F12/B	-1.67	2.02	2.36	6.0	64.C_3H_6
5T	UCCSD(T*)-INT-F12/B	-0.45	1.51	2.00	7.3	83.F_2O_2
5Q	UCCSD(T) -INT-F12/A	-0.18	1.11	1.58	5.1	83.F_2O_2
5Q	UCCSD(T) -INT-F12/B	-1.17	1.57	1.94	5.4	64.C_3H_6
5Q	UCCSD(T*)-INT-F12/B	0.12	1.33	1.88	7.2	83.F_2O_2

[a] Basis set combinations used for the additivity scheme. The first letter corresponds to the cardinal number of the SB, the second to the cardinal number of the LB.
For example, "TD" equals to the cc-pVTZ/cc-pVDZ-F12 basis set combination.

which achieve better or similar accuracy than the two "QT" models. These include a CCSD(T) calculation with either a quadruple- or quintuple-zeta quality basis sets and thus, making these models more computationally demanding than the previous. However, four of those achieve even higher accuracy than the one-basis protocol (Table 4.9) In particular, the UCCSD(T)-INT-F12/A model with the "5Q" combination of basis sets yields the lowest error deviations, with a MAD of 1.11 kJ/mol and a RMS value of 1.58 kJ/mol, again with respect to the ATcT reference atomization energies.

Comparison of all the models included in Table 4.10 with respect to the highly accurate results of Klopper *et al.* [114] gives comparable errors with those which were discussed in the previous paragraph. For example, the UCCSD(T)-INT-F12/A with the "QT" basis sets has an RMS of 1.32 kJ/mol, while for the most accurate of those models, the UCCSD(T)-INT-F12/A with the "5Q" basis has an RMS error less than 1 kJ/mol (0.92 kJ/mol).

Based on the statistical results showed in Table 4.10, conclusions can be made about a reasonable choice for an atomization energy protocol which includes the additivity scheme of Eq. (3.22). From the eleven models which have been discussed, five have RMS errors equal to or lower than 2 kJ/mol. From these five, two include the empirical scaling of (T*) and they are discarded, as far as they achieve similar accuracy with the non-empirical models. From the remaining three, two include a time-consuming CCSD(T)/cc-pV5Z step, while the fifth ("QT") uses the quadruple-zeta basis and provides almost the same accuracy (1.90 instead of 1.58 kJ/mol) with the "5Q" model. Thus, as a method of choice, the "QT" additivity scheme which is based on the UCCSD(T)-INT-F12 model with approximation A is suggested for a possible protocol.

It has already been mentioned that the accuracy of the additivity schemes which have been described in this Section is based partially on error cancellation. This is somehow true for the models which include energy components from small basis sets, like the cc-pVTZ/cc-pVTZ-F12 combination. The source of these errors arises from the basic principle of the composite scheme: the summation of individual terms. These terms tend to reach their complete basis set limit either as approximations (empirical or not) or as extrapolations but they also carry their individual

Table 4.11: Heats of formation with respect to H_2, CO, CO_2, N_2 and F_2 of 25 molecules from the 106-molecule Test Set (in kJ/mol).

	Reaction	Additivity Schemes				
		QZ-F12	TZ/TZ-F12	TZ/QZ-F12	QZ/TZ-F12	Reference[a]
1	$NCF+CO_2 \rightarrow 2CO + \frac{1}{2}N_2 + \frac{1}{2}F_2$	-181.5	-182.2	-181.5	-182.5	-181.2
2	$CNF+CO_2 \rightarrow 2CO + \frac{1}{2}N_2 + \frac{1}{2}F_2$	117.2	114.7	115.5	116.1	118.2
3	$CF_2+CO_2 \rightarrow 2CO + F_2$	-382.5	-383.2	-382.3	-383.7	-381.2
6	$CHF+CO_2 \rightarrow 2CO + \frac{1}{2}H_2 + \frac{1}{2}F_2$	-41.6	-41.9	-41.3	-42.3	-40.3
9	$NCH+CO_2 \rightarrow 2CO + \frac{1}{2}H_2 + \frac{1}{2}N_2$	-63.7	-63.8	-63.2	-64.1	-63.2
10	$CNH+CO_2 \rightarrow 2CO + \frac{1}{2}H_2 + \frac{1}{2}N_2$	-0.5	-0.8	-0.3	-1.0	0.2
15	$CH_2+CO_2 \rightarrow 2CO + H_2$	240.6	241.1	241.6	240.1	241.7
20	$CH_2O \rightarrow CO + H_2$	-23.2	-22.5	-22.2	-23.2	-22.3
21	$HCHO \rightarrow CO + H_2$	195.1	195.8	196.1	195.0	196.5
30	$CH_4+CO_2 \rightarrow 2CO + 2H_2$	-301.9	-300.2	-299.5	-302.3	-301.0
40	$C_2H_2+2CO_2 \rightarrow 4CO + H_2$	-156.8	-156.8	-155.6	-157.7	-154.9
73	$FH \rightarrow \frac{1}{2}H_2 + \frac{1}{2}F_2$	-283.1	-281.5	-281.7	-283.2	-283.4
74	$FHO+CO \rightarrow CO_2 + \frac{1}{2}H_2 + \frac{1}{2}F_2$	192.8	194.6	194.3	193.6	193.1
76	$FH_2N \rightarrow \frac{1}{2}F_2 + \frac{1}{2}N_2 + H_2$	-57.5	-55.0	-54.8	-57.0	-56.8
78	$FNO+CO \rightarrow CO_2 + \frac{1}{2}N_2 + \frac{1}{2}F_2$	204.5	201.8	202.0	204.6	206.1
82	$FO_2+CO \rightarrow CO_2+F_2$	316.4	315.8	316.0	316.8	318.1
85	$HNO+CO \rightarrow CO_2 + \frac{1}{2}H_2 + \frac{1}{2}N_2$	392.1	392.7	392.5	392.7	392.0
91	$H_2N_2(cis) \rightarrow H_2 + N_2$	195.2	195.2	195.6	195.1	195.4
92	$H_2N_2(trans) \rightarrow H_2 + N_2$	172.7	173.3	173.7	172.7	173.1
93	$H_2N_2(iso) \rightarrow H_2 + N_2$	275.7	277.0	277.0	276.2	276.9
95	$H_2O+CO \rightarrow CO_2+H_2$	29.7	32.5	31.9	30.4	29.6
96	$H_2O_2+2CO \rightarrow 2CO_2+H_2$	424.9	429.3	428.6	426.8	424.6
97	$NH_3 \rightarrow \frac{1}{2}N_2 + \frac{3}{2}H_2$	-82.1	-80.1	-80.0	-82.0	-81.9
102	$N_2O+CO \rightarrow CO_2+N_2$	369.0	366.3	366.4	368.8	368.9
105	$O_3+3CO \rightarrow 3CO_2$	1032.1	1027.4	1026.9	1033.5	1033.4
	Mean Error	0.71	0.48	0.23	0.71	
	MAD	0.77	1.79	1.52	1.07	
	RMS	0.95	2.33	2.17	1.32	

[a] From Ref [114].

errors. For example, the correlation energy of the CCSD(T)-INT-F12 method is a very good approximation to the CBS of the coupled-cluster method, but not the exact value. Another example is the empirical scaling of the perturbative triples which sometimes overestimates the CBS of the (T) energy component. The CABS singles, which are obtained from the explicitly-correlated MP2 calculation may again lead to overestimation of the HF energy. These errors are likely to cancel each other, especially in the schemes with the relative smaller basis sets. However, independent of the presence of error cancellation, the increase of the size of the basis sets of the additivity scheme decreases the respective atomization energy deviations. This is evident from the statistical results of Table 4.10. The schemes which include the quintuple-zeta basis set are much more accurate than the rest. In addition, increase of the size of the LB leads to further decrease of the mean absolute and RMS errors.

4.1.6 Reaction energies

The 106-molecule test set is also used for studying the heats of formation of 25 molecules with respect to the H_2, CO, CO_2, N_2 and F_2 molecules. The estimation of the heats of formation is straightforward as far as the total energies of these molecules have already been computed. A selection between the most accurate models which are based on the interference-corrected MP2-F12 was carried out and the applicability of those to the heats of formation has been examined. All individual results and statistics for each of these models are shown in Table 4.11. These correspond to the protocol where both CCSD(T) and second-order corrections have been calculated with the same basis set (cc-pVQZ-F12, column under the "QZ-F12" header) and the three different combinations of basis sets of the additivity scheme (cc-pVTZ/cc-pVTZ-F12, cc-pVTZ/cc-pVQZ-F12 and cc-pVQZ/cc-pVTZ-F12, shown in Table 4.11 under the headers "TZ/TZ-F12", "TZ/QZ-F12" and "QZ/TZ-F12", respectively). From the additivity schemes, the first two were chosen because they are not computationally demanding. Both of them also include the empirical (T*) scaling. On the contrary, the third one was chosen due to the accurate atomization energies which had been provided (see Section 4.1.5), without falling back on the empirical scaling. As reference, like in the previous analysis, the highly accurate results of Klopper et al. [114] were used. The same test set of the 25 heats of formation has been used from Köhn and Tew [109] on their article about the singles correlation contributions to the CCSD-F12 energy which reduce the basis set error in the Hartree-Fock energy.

From the four methods examined, the most accurate is the composite scheme which both CCSD(T) energy and second-order corrections from the interference-corrected MP2-F12 method are obtained from the same basis set (cc-pVQZ-F12). The very small deviations from the reference reaction energies have as a result a mean absolute deviation of 0.77 kJ/mol and a less than 1 kJ/mol RMS value (0.95 kJ/mol). Similar accuracy is achieved from the cc-pVQZ/cc-pVTZ-F12 additivity scheme, which includes the slightly less computational demanding step of CCSD(T)/cc-pVQZ. For most of the reactions, high accuracy has been achieved, but due to some outliers, like for example the heat of formation of difluoromethylene ($3.CF_2$) and acetylene ($40.C_2H_2$), the RMS value is slightly larger, at 1.32 kJ/mol. The additivity schemes which include the CCSD(T)/cc-pVTZ level of theory are about 1 kJ/mol less accurate, which RMS values larger than 2 kJ/mol. However, it should be kept in mind that these models avoid the calculation of the coupled-cluster energy with a quadruple-zeta basis set, avoiding in that manner this time-consuming step. On the contrary, part of the correlation energy is regained from the energy corrections of the interference-corrected MP2-F12 method.

4.2 AE6/BH6 Test Sets

Apart from the 106-molecule test set, alternative and more compact representative test sets exist in the literature. The six-membered atomization energy (AE6) and six-membered reaction barrier height (BH6) test sets of Lynch and Truhlar [127] are examples of such cases. These test sets have been proposed as representative sets for the 109 atomization energies and 44 barrier heights of the Database/3 data set [143]. Both sets contain a few but diverse cases and they pose a challenge for the protocols based on the interference-corrected MP2-F12 described in the previous Section.

The AE6 test set consists of the atomization energies of six molecules: SiH_4, SiO, S_2, propyne (C_3H_4), glyoxal ($C_2O_2H_2$) and cyclobutane (C_4H_8). The BH6 test set consists of the forward and reverse barrier heights of the three hydrogen transfer reactions $OH + CH_4 \rightarrow CH_3 + H_2O$, $H + OH \rightarrow O + H_2$, and $H + H_2S \rightarrow H_2 + SH$. The geometries used obtained at the QCISD/MG3 level of theory and they are supplied by Truhlar et al. [144].

Haunschild and Klopper [141] provided new theoretical reference values for AE6 and BH6 from explicitly-correlated coupled-cluster theory. Aim of that study was to propose an alternative computational protocol which comes close to the accuracy of the W4 method [145] and does not involve basis set extrapolations or any empirical factors.

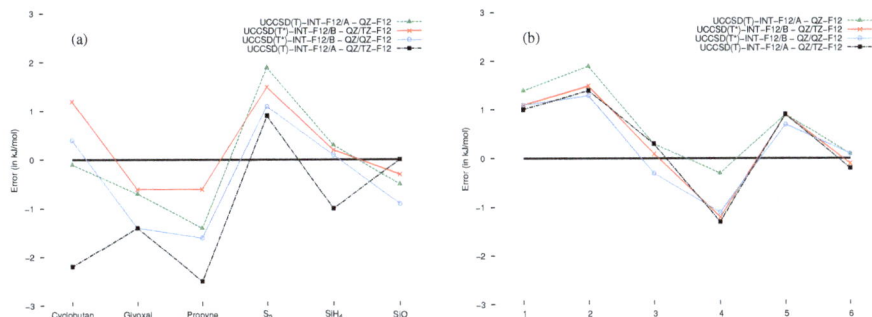

Figure 4.4: Energy deviations from the computational reference values of (a) AE6 and (b) BH6 test sets.

Reference values for the AE6 and BH6 test sets were obtained from that study. These values are at the CBS level of the fc-CCSD(T) method. The basis-set limit was calculated from Eq. (2.9). Apart from the reference values, two extra terms were also obtained from that study. These are the core/core-valence and higher-order terms (i.e. full triples and perturbative quadruples) which were added to the total energies of the CCSD(T)-INT-F12 method. The core/core-valence correction was computed from the difference between the all-electron and frozen-core CCSD(T)/cc-pwCVQZ calculations and the higher-order excitations from the difference between fc-CCSDT(Q) and fc-CCSD(T) calculations with the cc-pV(D+d)Z basis [146]. Scalar relativistic and spin-orbit effects are not considered in this study as far as the approximate CCSD(T) basis set limit is under examination, as it is reached with interference effects.

Both one-basis and additivity schemes were tested for their accuracy. For the scheme where all terms are calculated with the same basis set, the cc-pVQZ-F12 basis was chosen, due to its high accuracy in conjunction with the interference effects. For the latter approaches, the cc-pVQZ/cc-pVTZ-F12 and cc-pVQZ/cc-pVQZ-F12 basis set combinations were used. The choice made was based on the results of Section 4.1.5 which are summarized in Table 4.10. In particular, both basis set combinations were used with the UCCSD(T*)-INT-F12 method with approximation B, while the first was also used with the UCCSD(T)-INT-F12/A method.

Figure 4.4(a) includes the deviations of the atomization energies of the six molecules which compose the AE6 test set, as they have been computed from the four aforementioned methods. The "one-basis" scheme yields for four molecules errors smaller than 1 kJ/mol, but overestimates the atomization energy of propyne (-1.4 kJ/mol) and underestimates the energy of the S_2 molecule (1.9 kJ/mol). However, the mean absolute deviation of 0.82 kJ/mol and the corresponding RMS error (1.03 kJ/mol) is one more proof of the accuracy of this scheme. The same level of accuracy is achieved from the UCCSD(T*)-INT-F12/B scheme with the cc-pVQZ/cc-pVQZ-F12 basis set combination (RMS: 1.06 kJ/mol). The similar scheme which includes as LB a basis with one cardinal number less (cc-pVTZ-F12) than the previous combination yields even smaller errors. The RMS value of 0.87 kJ/mol, with a MAD of 0.73 kJ/mol, is the smallest from all the schemes tested with the AE6 set. On the contrary, the fourth scheme (UCCSD(T)-INT-F12/A with cc-pVQZ/cc-pVTZ-F12 basis set combination) has a MAD of 1.33 kJ/mol and a RMS of 1.57 kJ/mol. However, it's not only that two out of six molecules have a larger deviation than 2 kJ/mol, but also the fact that the errors of the atomization energies of the glyoxal and SiO molecules calculated from this specific scheme do not follow the same "trend" like they do for the other schemes. On the contrary, they have smaller errors than expected. The reason for than behavior may be a possible error additivity and cancellation between the different energy components which are summed up in this composite scheme.

The errors from the same methods for the six barrier heights of the BH6 test set are shown on Figure 4.4(b). The numbers 1 and 2 correspond to the forward and reverse barrier heights of the $OH + CH_4 \rightarrow CH_3 + H_2O$ reaction, numbers 3 and 4 to those of the $H + OH \rightarrow O + H_2$ reaction and numbers 5 and 6 to the $H + H_2S \rightarrow H_2 + SH$ reaction. All methods show almost the same behavior, having similar deviations from the reference values. The only exception is the fourth barrier height, where the one-basis scheme yields an error close to zero, while the other methods have a deviation of about 1 kJ/mol. For all reactions, these deviations are much smaller than those of the AE6 test set. They are lower than 2.0 kJ/mol, in absolute values, and the RMS errors are about 1.0 kJ/mol, for all four methods.

4.3 G2/97

The last collection of molecules on which the performance of the CCSD(T)-INT-F12 method has been tested is the G2/97 test set. [128, 129] The G2/97 test set is composed of 148 diverse molecules. It includes some difficult cases, like the carbonyl fluoride (COF_2) molecule, whose experimental reference value has a high uncertainty [147], 29 radicals, aromatic moieties, some relatively large molecules (e.g. benzene), plus molecules which are composed of atoms not included in the previous two sets (e.g. lithium, aluminum, sodium, etc.). All the members of the G2/97 test set are shown in Table 4.12.

Recently, new theoretical reference values have been published for the G2/97 test set. [10] These accurate values have been calculated from a computational protocol based on the explicitly-correlated CCSD(T) method. Corrections for higher excitations and core/core-valence correlation effects are accounted for. The main argument of the authors on why these theoretical reference values should be preferred over the existing experimental values is that the high uncertainty of experimental data makes the new reference more accurate on average. [10] Therefore, the effects of the interference effects on the CCSD(T) atomization energies of the 148 molecules will be compared with this new protocol.

The recalculation of the correlation components from higher excitations and core/core-valence effects are neglected in this Thesis. The reason for that is the fact that the accuracy of the CCSD(T)-INT-F12 method is examined and not a full thermochemical protocol. Thus, these energy terms have been subtracted from the reference values.[1] The reference explicitly-correlated CCSD(T) atomization energies are shown in the third column of Table 4.12. All geometries have been obtained from the B3LYP/6-31G($2df, p$) level of theory (from Ref. [148]).

The cc-pVQZ-F12 basis set was used for all calculations with the CCSD(T) method with second-order terms from the interference-corrected MP2-F12. The same computational procedure was followed for the AE6, BH6 and 106-molecule test sets, as it has been described in Section 4.1.2. The only difference is that the INT-MP2-F12 method was used with both invariant and fixed-amplitudes (sp cusp conditions) approaches and with either (semi-)canonical or with localized orbitals. Another, minor difference is that no aug-cc-pwCV5Z cbas exist for sodium, so the cc-pVQZ-F12 cbas was chosen for the calculations of the Na atom and the two molecules including it (Na_2 and NaCl).

---

[1]The individual correlation energy components have been kindly provided by Dr. Robin Haunschild, personal communication.

Table 4.12: Reference and calculated with the invariant CCSD(T)-INT-F12 method atomization energies of the G2/97 test set (in kJ/mol). The Boys localization scheme was used for the closed shell calculations. Values obtained with canonical orbitals are shown with *italic* font. For more details, see text.

Molecule		Ref	INT	Molecule		Ref	INT
$AlCl_3$	Aluminum trichloride	1307.5	1304.1	C_3H_4	Allene	2926.2	2925.5
AlF_3	Aluminum trifluoride	1802.9	1801.6	C_3H_4	Cyclopropene	2834.9	2834.4
BCl_3	Boron trichloride	1350.9	1348.2	C_3H_4	Propyne	2932.2	2931.6
BF_3	Boron trifluoride	1961.8	1960.2	C_3H_6	Cyclopropane	3554.4	3554.3
BeH	Beryllium monohydride	212.0	*209.5*	C_3H_6	Propene	3585.5	3585.0
CCl_4	Tetrachloromethane	1310.2	1309.9	C_3H_6O	Acetone	4077.8	4077.3
CF_4	Tetrafluoromethane	1997.1	1996.3	C_3H_7Cl	1-Chloropropane	4113.5	4113.1
CH	Methylidyne radical	350.3	*350.5*	C_3H_7	Isopropyl radical	3753.5	*3754.8*
CH_2Cl_2	Dichloromethane	1547.1	1546.3	C_3H_8O	Methoxyethane	4568.7	4568.8
CH_2F_2	Difluoromethane	1825.7	1825.2	C_3H_8O	Isopropyl alcohol	4622.2	4622.1
CH_2O_2	Formic acid	2089.5	2088.7	C_3H_8	Propane	4197.9	4197.8
CH_2O	Formaldehyde	1558.9	1558.3	C_3NH_3	Acrylonitrile	3169.1	3168.3
CH_2	Singlet carbene	754.3	754.4	C_3NH_9	Trimethylamine	4838.1	4838.7
CH_2	Triplet carbene	793.6	*793.7*	C_4H_{10}	Isobutane	5431.5	5431.5
CH_3Cl	Chloromethane	1649.7	1649.1	C_4H_{10}	n-Butane	5425.5	5425.5
CH_3	Methyl radical	1282.3	*1282.4*	C_4H_4O	Furan	4138.7	4138.2
CH_3O	Hydroxymethyl radical	1707.5	*1708.5*	C_4H_4S	Thiophene	4012.8	4012.1
CH_3O	Methoxy radical	1667.3	*1668.3*	C_4H_6	1,3-Butadiene	4214.2	4213.4
CH_3S	Methylthio radical	1593.9	*1595.2*	C_4H_6	2-Butyne	4177.3	4176.7
CH_4	Methane	1752.0	1752.0	C_4H_6	Bicyclo[1.1.0]butane	4103.5	4103.5
CH_4O	Methanol	2139.8	2139.5	C_4H_6	Cyclobutene	4167.8	4167.4
CH_4S	Thiomethanol	1979.8	1979.3	C_4H_6	Methylenecyclopropane	4131.6	4131.3
$CHCl_3$	Trichloromethane	1436.4	1435.6	C_4H_8	Cyclobutane	4790.0	4790.1
CHF_3	Trifluoromethane	1916.8	1916.2	C_4H_8	Isobutene	4827.1	4826.8
CHO	Formyl radical	1160.9	*1161.5*	C_4H_9	*tert*-Butyl radical	4994.7	*4997.4*
CN	Cyano radical	742.9	*743.6*	C_4NH_5	Pyrrole	4461.4	4461.3
CNH	Hydrogen cyanide	1300.4	1299.9	C_5H_8	Spiropentane	5348.7	5348.8
CNH_3O_2	Methyl nitrite	2492.2	2492.8	C_5NH_5	Pyridine	5152.3	5152.1
CNH_3O_2	Nitromethane	2501.7	2501.9	C_6H_6	Benzene	5693.9	5693.5
CNH_5	Methylamine	2425.5	2425.6	Cl_2	Dichlorine	245.9	245.7
C_2Cl_4	Tetrachloroethylene	1960.5	1958.0	CO	Carbon monoxide	1080.4	1079.7
C_2F_4	Tetrafluoroethylene	2453.7	2452.9	CO_2	Carbon dioxide	1621.6	1620.3
C_2H	Ethynyl radical	1097.8	*1097.9*	COF_2	Carbonyl fluoride	1751.3	1750.2
C_2H_2	Acetylene	1683.2	1682.6	COS	Carbonyl sulfide	1394.2	1392.9
$C_2H_2O_2$	Glyoxal	2640.4	2639.3	CS	Carbon monosulfide	712.5	711.3
C_2H_2O	Ketene	2216.2	2215.3	CS_2	Carbon disulfide	1161.4	1159.7
C_2H_3Cl	Vinyl chloride	2263.7	2262.8	FCl	Chlorine monofluoride	259.7	259.7
C_2H_3	Vinyl radical	1852.4	*1853.0*	F_2	Difluorine	158.2	158.7
C_2H_3F	Vinyl fluoride	2387.5	2386.9	F_3Cl	Chlorine trifluoride	529.3	530.2
C_2H_3O	Carbonyl methane	2422.4	*2424.5*	HCl	Hydrogen chloride	448.2	447.3
C_2H_3OCl	Acetyl chloride	2783.4	2782.3	HF	Hydrogen fluoride	591.3	590.9
C_2H_3OF	Acetyl fluoride	2942.6	2941.8	HOCl	Hypochlorous acid	691.2	691.1
C_2H_4	Ethylene	2347.1	2346.7	HO	Hydroxyl radical	446.8	*446.8*
$C_2H_4O_2$	Acetic acid	3346.9	3346.3	HS	Mercapto radical	365.2	*365.2*
$C_2H_4O_2$	Methyl formate	3276.4	3275.8	H_2	Dihydrogen	457.7	457.9
C_2H_4O	Acetaldehyde	2821.3	2820.7	H_2O_2	Hydrogen peroxide	1120.6	1120.7
C_2H_4O	Oxirane	2712.3	2712.1	H_2O	Water	971.8	971.4
C_2H_4S	Thiirane	2607.2	2606.9	H_2S	Hydrogen sulfide	766.2	765.5
C_2H_5Cl	Ethyl chloride	2885.7	2885.2	LiF	Lithium fluoride	581.8	581.6
C_2H_5	Ethyl radical	2515.3	*2515.9*	LiH	Lithium hydride	242.3	242.5
C_2H_5O	Ethoxy radical	2906.3	*2908.7*	Li_2	Dilithium	101.2	101.4
C_2H_6	Ethane	2970.5	2970.5	Na_2	Disodium	71.6	68.7
C_2H_6O	Dimethyl ether	3328.5	3328.5	NaCl	Sodium chloride	411.6	410.9
C_2H_6O	Ethanol	3379.5	3379.3	NF_3	Trifluoroamine	856.4	857.2
C_2H_6OS	Dimethyl sulfoxide	3570.7	3570.4	NH_2	Amino radical	760.2	*760.4*
C_2H_6S	Dimethyl sulfide	3204.3	3204.1	NH_3	Ammonia	1241.8	1241.7
C_2H_6S	Thioethanol	3209.0	3208.6	NH	Imidogen	345.8	*346.1*
C_2N_2	Cyanogen	2078.2	2077.2	NO_2	Nitrogen dioxide	940.8	*942.8*
C_2NF_3	Trifluoroacetonitril	2669.9	2668.8	NOCl	Nitrosyl chloride	791.4	791.2
C_2NH_3	Acetonitrile	2562.7	2562.2	NO	Nitric oxide	632.2	*633.0*
C_2NH_5	Aziridine	2998.4	2998.6	N_2	Dinitrogen	947.8	947.4
C_2NH_5O	Acetamide	3615.0	3614.7	N_2H_4	Hydrazine	1824.8	1825.1
C_2NH_7	Dimethylamine	3625.9	3626.2	N_2O	Nitrous oxide	1120.4	1120.0
C_2NH_7	Ethylamine	3660.1	3660.2	OCl	Monochlorine monoxide	266.4	*268.2*

Table 4.12: *(Continued.)*

Molecule		Ref	INT		Molecule		Ref	INT
OF_2	Difluorine monoxide	385.6	386.5		$SiCH_6$	Methylsilane	2625.1	2624.9
OS	Sulfur monoxide	522.4	*523.2*		$SiCl_4$	Silicon tetrachloride	1624.0	1621.1
O_2	Dioxygen	499.8	*500.1*		SiF_4	Silicon tetrafluoride	2413.4	2411.8
O_2S	Sulfur dioxide	1079.8	1078.7		SiH_2	Singlet silylene	642.4	642.4
O_3	Ozone	598.7	599.4		SiH_2	Triplet silylene	557.1	*557.1*
P_2	Diphosphorus	481.2	480.4		SiH_3	Silyl radical	953.8	*954.0*
PF_3	Phosphorus trifluoride	1524.4	1524.1		SiH_4	Silane	1357.6	1357.4
PH_2	Phosphino radical	643.3	*643.5*		SiO	Silicon monoxide	800.3	799.8
PH_3	Phosphane	1009.3	1009.2		Si_2H_6	Disilane	2239.1	2238.7
S_2	Disulfur	428.6	*428.6*		Si_2	Disilicon	301.4	*301.4*

Table 4.13 includes the statistical errors from all the different schemes used for the calculation of the atomization energies of the G2/97 test set. For the invariant CCSD(T)-INT-F12 method with the (semi-)canonical orbitals, deviations from the reference values of Haunschild and Klopper are higher than it was expected based on the results obtained from the one-basis approach for the other test sets. The MAD of 1.8 kJ/mol dictates to investigate the specific approach. Results from the fixed-amplitude interference-corrected MP2-F12 method are similar but slightly more accurate than those from the invariant scheme. In general, the fixed-amplitudes atomization energies differ about 0.3 to 0.5 kJ/mol from the reference values. This holds for most of the molecules which have large errors from invariant INT-MP2-F12. Although this is a small correction for most of the members of the G2/97 test set, the MAD of the fixed-amplitudes approach still remains relatively large, at 1.68 kJ/mol. The reason that the MAD value remains outside of the initial target of the CCSD(T)-INT-F12 (i.e. the sub-chemical accuracy) is the existence of a few cases where the fixed amplitudes introduce larger errors, like the tetrachloromethane (-4.5 instead of -3.9 kJ/mol) and the tetrafluoromethane (-4.6 instead of -2.4 kJ/mol) molecules.

Apart from the differences between the invariant and the fixed-amplitudes, specific cases show exceptionally high deviations and raise the RMS values above 2 kJ/mol for *both* fixed and invariant approaches. Most of them have one thing in common: there are non linear molecules with high symmetry, like the chlorine trifluoride (F_3Cl), which has the maximum error of -7.8 kJ/mol for this test set, the trichloromethane ($CHCl_3$, -5.4 kJ/mol) and the trifluoroamine (NF_3, -5.0 kJ/mol) molecules, or they hold one highly symmetrical functional group, like, for example, the CF_3 group of the trifluoroamine (NF_3, -5.0 kJ/mol) and trifluoroacetonitril (C_2NF_3, -5.9 kJ/mol) molecules. Some aromatic molecules show also notable deviations, like pyridine (-4.0 kJ/mol) and furan (-3.8 kJ/mol). These large deviations originate from the degeneracies or near-degeneracies of these molecules.

Till now, in all the above cases SCF (semi-)canonical orbitals with symmetry restrictions were used. These starting orbitals provided good accuracy for the previous test sets examined. However, it was already pointed out that specific cases need extra caution (Section 3.3.4). The G2/97 test set includes such cases, much more than the previous sets included. Thus, in order to reduce such errors introduced from these cases and correct the calculated atomization energies, the Boys localized orbitals are applied in the INT-MP2-F12 method. This has been done for the closed shell molecules of the G2/97 test set. Table 4.12 includes all these results, next to the reference values, for the 148 members of the test set. The current status of the code does not support localization schemes for alpha and beta orbitals of open-shell calculations. Thus, canonical orbitals have still been used for the thirty open-shell molecules out of 148 and their atomization energies are shown with *italic* font. However, this does not affect significantly the final MAD which is lowered from

Table 4.13: Statistical errors for the G2/97 test set (in kJ/mol).

		Mean Error	MAD	RMS	Max Error	Molecule
Canonical	Fixed	-1.58	1.68	2.18	-6.7	F_3Cl
	Inv	-1.70	1.80	2.35	-7.8	F_3Cl
Localized	Fixed	0.50	0.84	1.16	4.8	$SiCl_4$
	Inv	0.21	0.65	0.94	3.4	$AlCl_3$

1.80 to 0.65 kJ/mol (Table 4.13). Most of these 30 open-shell moieties have very small deviations and only in four cases (beryllium monohydride, carbonyl methane, ethoxy radical and *tert*-butyl radical) the error exceeds 2 kJ/mol. Indeed, if the thirty open-shell cases are excluded from the statistical analysis, the mean, MAD and RMS values are 0.42, 0.61 and 0.88 kJ/mol, respectively, which are almost identical to those for the whole test set. It should also be noted that the invariant MP2-F12 calculation of the tetrachloroethylene molecule (C_2Cl_4) had difficulties to converge (the B matrix was not positive definite). In Table 4.12, the atomization energy computed with the fixed-amplitude ansatz is shown and, for sake of consistency, its deviation was excluded from the statistical analysis of Table 4.13.

The localized MOs have corrected the large errors (till -7.8 kJ/mol) which have been reported for the CCSD(T)-INT-MP2 method with the symmetrical (semi-)canonical orbitals. For most of these cases, the deviations have dropped below 1 kJ/mol. In general, the MAD is dropped to one third of the description with the canonical MOs and, interestingly, the RMS value to below 1 kJ/mol (0.94 kJ/mol). The fixed-amplitudes approach with LMOs achieves also similar accuracy (RMS of 1.16 kJ/mol).

The only cases that still show errors above 1.5 kJ/mol are a few open-shell systems, the $AlCl_3$, BCl_3, $SiCl_4$ molecules and the two molecules that contain the sodium atom (Na_2 and $NaCl$). A possible source for the errors of the remaining closed shell members of the G2/97 test set may be the total energies of the atoms. In these energies, the δE_{INT} terms have been calculated with the (semi-)canonical orbitals. The error introduced is negligible for the relatively "small" atoms (like C, N, O anf F), but it may be important for the larger, like Na, Al and Cl. This also explains the good accuracy achieved for the 106-molecule test set, where such atoms were absent.

4.4 Summary

In this chapter, the applicability of the coupled-cluster singles-doubles and perturbative triples with second-order corrections from the interference-corrected explicitly-correlated MP2 was tested for obtaining highly accurate atomization energies and reaction barrier heights. Two different approaches were tested; the first was based on the use of the same basis set for the calculation of the individual terms of the CCSD(T)-INT-F12 method ("one-basis") while the second, named as additivity scheme, was obtaining different energy components from different basis sets. For both approaches, their correlation energy was firstly compared with the explicitly-correlated CCSD(T) correlation energy in order to choose between different basis sets and F12 approximations (A, B or averaged 60% A - 40% B). The choice was based on two factors: to obtain high accuracy by keeping as low computational cost as possible. Based on this comparison, the UCCSD(T)-INT-F12/A model with the cc-pVQZ-F12 basis was proved to be the best option for the one-basis approach. For the additivity schemes, four models were chosen for providing mean absolute deviations below 1 kcal/mol, without including large basis sets.

In addition, the performance of the empirical scaling of the perturbative triples, which is also known as (T*), was discussed. The (T*) correction proved to be important only when it is combined with schemes with small and medium basis set sizes, while for larger basis sets, the results were less accurate.

By taking into account the above considerations, composite models based on the INT-MP2-F12 were tested with respect to the atomization energies of a test set composed of 106 molecules. Extra corrections were added to these models, like the core/core-valence correlation correction, the CABS singles correction, higher excitations corrections, scalar relativistic effects, spin-orbit interactions for the atoms, and anharmonic zero-point vibrational energies. Some of these models provided very accurate atomization energies like, for example, the UCCSD(T)-INT-F12/A model with the cc-pVQZ-F12 basis which has a MAD of 1.31 kJ/mol and the UCCSD(T)-INT-F12/A model based on the additivity scheme with the cc-pVQZ/cc-pVTZ-F12 basis set combination which has a MAD of 1.47 kJ/mol. Highly accurate heats of formation with respect to H_2, CO, CO_2, N_2 and F_2 of 25 molecules were also obtained from these two models.

The AE6 and BH6 test sets of Lynch and Truhlar were also computed from the computational models proposed in this chapter. The RMS values of around 1 kJ/mol for both test sets indicate that accurate model chemistries based on the interference-corrected MP2-F12 method can be formulated.

Finally, in Section 4.3 the G2/97 test set was studied. G2/97 includes more challenging and difficult cases than the previous test sets like radicals (open-shell) molecules, aromatic molecules and molecules containing atoms not present in the previous sets (i.e. sodium, aluminum, beryllium etc.). Excellent agreement with the CCSD(T)(F12)/cc-pVQZ-F12 level of theory is achieved with the one-basis approach when localized orbitals are used (for the closed shell cases) in conjunction with interference-effects. For the invariant CCSD(T)-INT-F12/cc-pVQZ-F12 level, the corresponding MAD is 0.65 kJ/mol and the RMS value below 1 kJ/mol. The conclusions reached from the analysis of the data obtained for the G2/97 test set can also be used in order to achieve even higher accuracy from a thermochemistry protocol based on the advantages of the CCSD(T)-INT-F12 method.

CHAPTER 5 ■

Noncovalent Interactions

5.1 Introduction

Modern chemistry is based on the understanding of the chemical bond. The chemical bond implies the distribution and delocalization of electrons over the entire molecule resulting in a strong, i.e., covalent interaction. The covalent description is fully adequate when the molecule is considered in free space, i.e., isolated from its surroundings. Modern theoretical *ab initio* quantum chemistry methods have been extremely successful in describing the electronic structure of isolated molecules to a degree of precision that in some cases comes very close to high-resolution spectroscopic results.

Atoms and molecules can interact together leading to the formation of either a new molecule or a molecular cluster. The former is clearly a covalent interaction; the latter one in which a covalent bond is neither formed nor broken is termed a noncovalent or van der Waals interaction. [149] The origin of the covalent and the noncovalent interactions is not their only difference. Covalent interactions are of short range and covalent bonds are generally shorter than 2 Å while noncovalent interactions are known to act at distances of several angstroms. The total stabilization energy of a molecular cluster lies usually between 1 and 20 kcal/mol, considerably smaller than the binding energy of a covalent bond of about 100 kcal/mol.

While covalent interactions determine the primary structure of a molecule, the noncovalent (weak) interactions are responsible for the tertiary and quaternary structure of a molecule and thus create the fascinating world of the 3D architectures of biomolecules. [150] For example, the double helical structure of DNA is of fundamental importance for its function: it allows it to store and transfer genetic information. The structure of the macromolecule should be rigid to maintain the double helix and floppy to allow for its opening. Very strong covalent bonds cannot fulfill these criteria, but noncovalent interactions, which are about 2 orders of magnitude weaker, can. Another example is the interaction of a ligand with a protein target, which is fully determined by various types of noncovalent interactions and solvation/desolvation processes, represents the key step in *in silico* drug design, which is nowadays a key step in the development of new potent drugs.

Noncovalent interactions are also responsible for gas separation from porous materials, like zeolites, carbon-based materials and metal-organic frameworks (MOFs). These host/guest interactions between the walls of the pores and the gas molecules play a vital role in the selective sorption of one gas over another. [151] A typical example of such processes is the carbon dioxide separation and storage from zeolites [152], activated carbon porous materials [153] and MOFs [154]. Another process which has attracted considerable attention is the hydrogen storage [155] for potential usage as motor fuel for automobiles [156]. The different parameters controlling the selective gas separations, like the size and composition of the framework pores or the properties of the guest molecules (kinetic diameter, high dipole or quadrupole moment, etc.) constitute a fascinating active scientific field with industrial applications. Taylor-made materials can be synthesized or functionalized with post-synthetic methods [157] according to our needs.

Last, but not least, noncovalent interactions are playing a key role in the spatial organization of the supramolecular assemblies. These stabilization forces of the host/guest complexes are mainly hydrogen bonds, stacking interactions, electrostatic interactions, hydrophobic interactions, charge-transfer interactions, and metal coordination. [158] The concept of self-assembly is grounded on such weak bonding between two molecules (intermolecular) or from the same molecule (intramolec-ular), which is formed in a suitable environment. Supramolecular chemistry has been applied in many and crucial applications, like, for example, in material chemistry, in catalysis and in medicine.

Although density functional theory (DFT) is widely used for obtaining geometries, energies and possibly the forces of these systems, it has its limitations. In particular, DFT does not recover the attractive dispersion (London) forces between molecules, although methods have been proposed in which an empirical dispersion term is added [159–162]. Dispersion forces and aromatic inter-actions pose in general a challenge for modern wave function and DFT methods. [163] Another alternative is to adopt a well-established procedure of proposing or calculating a potential energy function that describes the energy of the system as a function of the positions of the atoms and then using this to study the motion of the system by classical mechanics (molecular dynamics) or to explore the potential surface to find the energy minima and the pathways between them. [164] Typically, these empirical potentials are based on point charges, but it has also been recognized that this is inadequate, especially in hydrogen-bonded systems. [165] Distributed multipole analy-sis [166] derives a description in terms of charges, dipoles and so on from a calculated wave function.

It is widely known that one of the most accurate descriptions of noncovalent bonds (in the range of chemical accuracy) is obtained from coupled-cluster calculations covering the singles and doubles excitations iteratively and the triple electron excitations pertubatively (CCSD(T)) at the com-plete basis set (CBS) limit. [167] Thus, the CCSD(T)/CBS level of theory provides benchmark data which can be used either for investigating the nature of noncovalent interactions in various binding motifs or for testing and/or parametrizing other, more computationally economical wave function or density functional theories as well as semiempirical methods and empirical potentials. Additionally, the CCSD(T)/CBS limit is adequate for the quantitative description of the noncova-lent interactions and higher-order energy contributions (i.e. full triples or pertubative quadruple excitations) can be neglected.

The development of faster quantum mechanical procedures is highly important since larger frag-ments of bio- and nanostructures need to be described as accurately as possible. It is also clear that interactions within these systems are governed not only by classical short- and long-range interactions, which can be basically described by empirical potentials, but also by effects that are clearly of quantum origin. Fast and accurate quantum mechanical methods are also needed in molecular dynamic simulations where the description of quantum effects plays a decisive role.

This Chapter is pointing exactly at this direction, i.e. the investigation of a computational faster quantum chemistry method, without any loss of accuracy. The basis set limit of the CCSD(T) model is approached faster with extrapolation schemes, when explicitly-correlation is used or when lower-order corrections are added to the total energy. The first two approaches have been dis-cussed in Chapter 2 while the third is one of the central ideas behind the new method presented in this Thesis. Interference-corrected second-order perturbation theory provides the tools for a faster convergence to the basis set limit, as it has been pointed out in Chapter 3. In this Chapter, the applicability of this method (CCSD(T)-INT-F12) will be examined in various weakly interacting systems, like noble gas dimers, dispersion-dominated complexes, a benchmark database specialized for studying noncovalent interactions, interactions between aromatic moieties and hydrogen adsorption in metal clusters..

5.1.1 Theoretical background

The approach of second-order corrections to the CCSD(T) energy has been used successfully for describing weak interaction energies by Hobza and co-workers. [124–126, 150, 167–169] The hybrid CCSD(T) basis set limit was obtained from:

$$E_{CCSD(T)/CBS} \approx E_{HF} + \delta E_{MP2/CBS} + \Delta E_{CCSD(T)}, \tag{5.1}$$

where the MP2 correlation energy is extrapolated to the CBS limit using Helgaker's formula [31] from the aug-cc-pVTZ and aug-cc-pVQZ basis sets and the HF energy was obtained from the calculation with the larger quadruple-zeta quality basis. The $\Delta E_{CCSD(T)}$ term corresponds to the difference between the CCSD(T) and MP2 correlation energies, as they are evaluated with a smaller basis set (aug-cc-pVDZ).

In particular, Hobza and co-workers have obtain from Eq. (5.1) reference interaction energies of small model complexes, DNA base pairs and amino acid pairs and they proposed two databases containing 22 [124] and 66 [168] complexes named as S22 and S66, respectively. These benchmark databases are composed of systems covering the most common types of noncovalent interactions in biomolecules, while keeping a balanced representation of dispersion and electrostatic contributions. Note that Eq. (5.1) is equivalent to Eq. (3.21), which corresponds to the unscaled case of δE_{INT} correction component, (i.e. the interference factors for each electron pair are equal to one).

A vital concern which must be considered for a successful description of a noncovalent bonded system is the treatment of the basis set superposition error (BSSE). This error is attributed to the incompleteness of the AO basis set. The most straightforward way to alleviate artificial stabilization due to the use of unequally large basis sets for the dimer and monomers is by applying the counterpoise (CP) correction as formulated by Boys and Bernardi [170], i.e. by using the same AO basis set for the monomers and the dimer. The counterpoise corrected energy E_{AB}^{cp} can be defined as

$$E_{AB}^{cp} = E_{AB}(AB) - E_{AB}(A) - E_{AB}(B) + E_A(A) + E_B(B), \tag{5.2}$$

where $E_N(M)$ is the energy of the molecular system M in the basis N. The notation N=AB means that the respective molecular system is computed using all of the basis functions of the whole complex AB. For example, $E_{AB}(A)$ is the energy of the fragment A obtained from a calculation in the basis of complex AB.

The interaction energy between the monomers of these two complexes can be computed either without (w/o CP; ΔE_{int}) or with applying a counterpoise correction (w/ CP; ΔE_{int}^{cp}). [1] These interaction energies are defined as follows:

$$\text{w/o CP:} \quad \Delta E_2 = E_{AB}(AB) - E_A(A) - E_B(B); \tag{5.3}$$

$$\text{w/ CP:} \quad \Delta E_2^{cp} = E_{AB}^{cp}(AB) - E_A(A) - E_B(B). \tag{5.4}$$

---

[1] Note that the term *int* with small letters stands for interaction energy and not for the interference correction δE_{INT}.

Note that the interaction energies ΔE_2 and ΔE_2^{cp} are two-body terms that are defined by dissociating the complex into its individual fragments A and B, while keeping the geometries of these fragments fixed. The one-body term

$$\Delta E_1 = E_A(A) - E_A^{eq}(A) + E_B(B) - E_B^{eq}(B) \tag{5.5}$$

accounts for the relaxation of the fragment geometries, where E_N^{eq} is the energy of the molecular system M in its own equilibrium geometry. Finally, the counterpoise-corrected and not-counterpoise-corrected binding energies are computed as $D_e^{cp} = -\Delta E_2^{cp} - \Delta E_1$ and $D_e = \Delta E_2 - \Delta E_1$, respectively.

5.1.2 Interactions between CO_2 and N-containing heterocycles

The author of this Thesis has also used in the past [11, 171] this approach for a study on the interaction between the carbon dioxide molecule and nitrogen-containing heterocycles. The basis set limit of MP2 was calculated with the explicitly-correlated MP2 method and the $\Delta E_{CCSD(T)}$ term with the 6-311++G** basis. Interaction energies obtained from this scheme were compared with two different spin-component-scaled MP2 approaches, denoted SCS-MP2 [115] and SOS-MP2 [172], and with a variety of DFT functionals with the addition of an empirical dispersion correction of DFT-D2 [160]. The acronyms SCS and SOS stand for spin-component-scaled and scaled-opposite-spin, respectively.

Starting point of that study was the carbon dioxide storage in MOFs and in particular in a sub-category called zeolitic imidazole frameworks (ZIFs). Their secondary building units (SBUs) are typically composed of zinc(II) or cobalt(II) and their organic part of imidazolate/imidazolate-type linkers. One of the main attributes of ZIFs is their zeolitic topology. The reason for this occurrence is that the metal-imidazole-metal angle is close to 145°, which is coincident with the Si-O-Si angle of the natural zeolites. These materials also share many advantages of the zeolitic chemistry, plus their exceptional thermal and chemical stability. [173]

The theoretical approach to this issue had two directions. The first was to examine the nature of the noncovalent forces and the properties of the pyridine-CO_2 van der Waals complex. The CCSD(T)/CBS best estimation of the binding energy (4.46 kcal/mol) supports the notion that the interaction between the two compounds is weak. The conclusions from the electronic structure theory were confirmed experimentally by Doran et al.. [174] The structure of the pyridine-CO_2 complex was analyzed by rotational spectroscopy using a Fourier transform microwave spectrometer. [175] The C_{2v} planar, vibrationally averaged structure established experimentally is in excellent agreement with the lowest-energy equilibrium structure calculated at the MP2/aug-cc-pVTZ level of theory and compared with seven energetically higher stationary points. The experimental value of the $C\cdots N$ van der Waals bond distance (2.7977(64) Å) is only 0.024 Å longer than the computational derived value of 2.774 Å. Finally, the calculated barrier of internal rotation of the CO_2 moiety is 1.2 kcal/mol and consistent with the lack of observable internal rotation from the experiment.

The second direction corresponded to a comparison of the binding energy of the CO_2 molecule on a series of nitrogen-containing heteroaromatic compounds and the corresponding physical phenomena which are stabilizing each of these systems. The purpose of this comparison was to suggest polar organic groups or MOF linkers with higher adsorption of CO_2. In this study a higher binding energy was observed for adenine and imidazopyridamine.

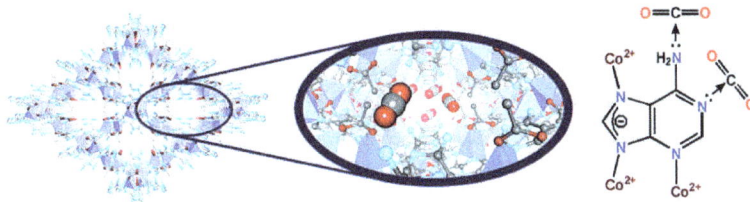

Figure 5.1: Pyrimidine and amino groups on the walls of the pores of bio-MOF-11 offer higher adsorption and selectivity of CO_2 over N_2. "Reprinted with permission from Ref. [181]. Copyright 2011 American Chemical Society."

Similar computational investigations of carbon dioxide complexes have been carried out. The interactions of small molecules, including CO_2, with two pyrazine molecules, used as organic linkers in MOFs were studied with post-HF methods (MP2, MP4(SDQ), MP2.5 [176] CCSD(T)) [177]. de Lange and Lane [178, 179] have studied with the CCSD(T)-F12 method small CO_2 dimers and cooperative many-body complexes which exhibit simultaneous CO_2-Lewis acid and CO_2-Lewis base intermolecular interactions. A DFT-D study [180] predicted that nitrogen doping of "zigzag" single-walled carbon nanotubes increases the binding of CO_2 by approximately 3 kcal/mol.

By following these ideas, new, advantageous porous materials which include on their pores nitrogen-containing heteroaromatic linkers can be designed *in silico* [182], synthesized [181, 183–186] or post-synthetically modified [187–189], fulfilling the desirable target of higher and selective CO_2 adsorption. For example, Rosi *et al.* [181] reported higher CO_2 capacity and exceptional selectivity for CO_2 over N_2 at 273 K in a material (bio-MOF-11) which has its pores decorated with pyrimidine and amino groups (Figure 5.1). Adenine has multiple Lewis basic sites, including an amino group and pyrimidine nitrogens, which can (weakly) interact with CO_2, resulting materials with high CO_2 adsorption energies. Thus, they propose that adenine-based MOFs would be ideal for selective CO_2 adsorption.

This Chapter has been divided into four parts. Firstly, a short discussion about the applicability of the CBS model chemistries for noncovalent interactions will be given and some simple test cases will be analyzed. (Section 5.2). In Section 5.3 results on the S22 benchmark database [126] from the CCSD(T)-INT-F12 method will be given. and will be compared with other accurate methods from the recent literature. In Section 5.4 two T-shape complexes of benzene with the imidazole and pyrrole aromatic moieties will be discussed. Finally, noncovalent interactions between the hydrogen molecule and exposed (undercoordinated) metal sites will be mentioned in Section 5.5

5.2 Interference Effects and Noncovalent Interactions

5.2.1 CBS model chemistries and noncovalent interactions

The central equation of the interference-corrected explicitly-correlated MP2 method (Eq. (3.1)) can be interpreted as extension to Eq. (5.1) described above. Main difference is that the $\delta E_{MP2/CBS}$ and δE_{MP2} terms are scaled from the interference factors for every second-order pair energy difference. This corrects the basis set truncation error of the higher-order CCSD(T) energy, as it has been discussed in the previous chapters.

The CBS family of model chemistries of Petersson and co-workers [38, 43–49], which is the precursor of the interference-corrected MP2-F12 method, have not been used thoroughly for the study of noncovalent interactions, apart from a few exceptions [190–192]. In 1998, Rablen *et al.* [190] have examined a set of 53 hydrogen-bonded complexes of water with various small organic molecules. In their study, binding energies calculated with MP2 and DFT with the B3LYP functional have been compared with the CBS-4 and CBS-Q model chemistries [46]. The main differences between these two protocols are the geometry optimization method (UHF for CBS-4, MP2 for CBS-Q), the basis sets used for every energy term (CBS-Q uses always larger basis sets that CBS-4) and the highest correlation energy component that is added (MP4(SDQ) for CBS-4, QCISD(T) for CBS-Q). Based on their analysis, they concluded that CBS-4 is less than optimal due to the underestimation of the binding energies. The reason may be either the insufficient method applied for the geometry optimization step used in the specific protocol or the lack of the important triples excitations which are missing in the MP4(SDQ) energy term. For example, for the $CH_3Cl \cdots H_2O$ complex, CBS-4 yielded an interaction energy of 1.21 kcal/mol, CBS-Q 3.31 kcal/mol, while B3LYP and MP2 resulted 2.80 and 4.60 kcal/mol, respectively. Both B3LYP and MP2 account for the counterpoise (CP) correction for the BSSE. For the majority of the systems, the B3LYP results with the CP correction are very close to those from the CBS-Q protocol. However, even if the B3LYP functional is used nowadays successfully in routine calculations, the specific study can not certify the applicability of the CBS-Q model chemistry for the description of noncovalent interactions.

Mascal *et al.* [191] have studied the noncovalent bonding between anions and the aryl centroid of electron-deficient aromatic rings, like 1,3,5-triazine, at the MP2 theory. Two representative complexes out of the 16 which were included in their publication were chosen for evaluation with the CBS-Q protocol. They reported that even if the structures of the optimized complexes changed very little, the interaction energies for both complexes fall between the uncorrected and CP-corrected MP2 values. Their argument that CP-corrected MP2 usually overestimates noncovalent interactions is reasonable, but again the study does not provide an argument for the success of the CBS-Q model chemistry.

A more detailed study of the description of weak interactions was carried out by Zhao and Truhlar [192]. They tested a series of different model chemistries and DFT functionals (in total 57 methods), including the CBS-QB3, which differs from CBS-Q only in the localization scheme used. The complexes studied are divided into five databases, depending on the various kinds of noncovalent interactions: six hydrogen bonding dimers, seven charge-transfer complexes, six dipole interaction complexes, seven "weak" interaction complexes (mainly composed of noble gases) and five $\pi - \pi$ stacking complexes. The mean absolute deviations (MAD) of CBS-QB3 for these databases are 0.17, 0.38, 0.36, 0.11, 0.57 and 0.32 kcal/mol, respectively. Even if the specific model chemistry is considered as adequately accurate, the test sets do not include any relatively difficult case. An additional important difference from the original CBS-QB3 protocol is the geometry optimization step which was skipped for the specific study. Instead of the MP2/6-31G† geometries [38, 43], the MC-QCISD/3 [143] level of theory was used.

5.2.2 Noble gas dimers

As a rule of thumb, a method which can accurately describe complexes which are interacting very weakly, can also be used for the study of any chemical system with noncovalent bonds. Typical complexes with weakly intermolecular interactions are the noble gas dimers. Thus, as a first step, the helium, neon and argon dimers have been studied and are analyzed in this Section. However, it should be kept in mind that the noble gas dimers as typical models for dispersion are rather poor representatives for dispersion interactions in large molecules. [126] Additionally, very accurate calculations with conventional CCSD(T) and different variants of explicitly-correlated CCSD(T)

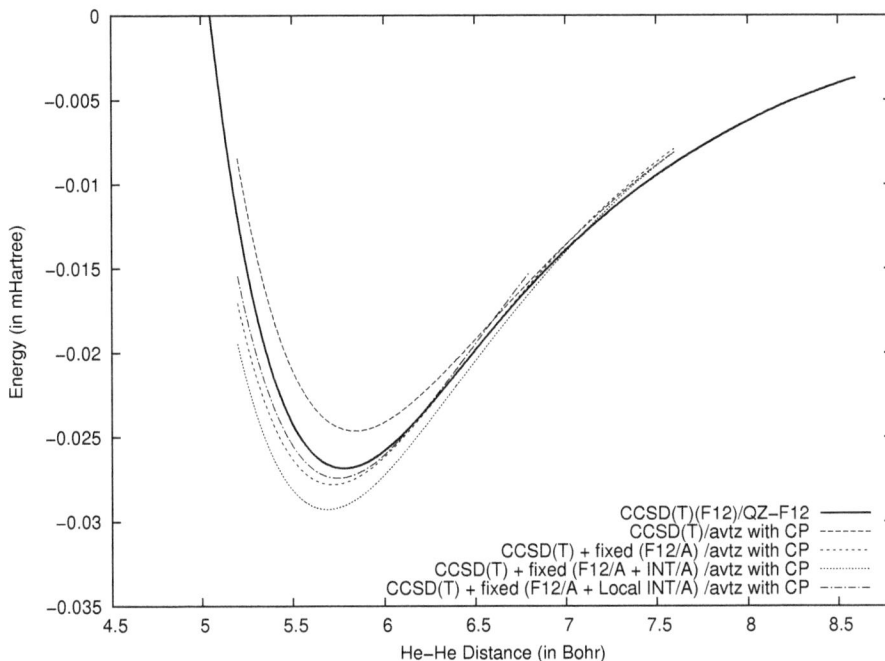

Figure 5.2: Potential energy curve of the helium dimer.

theories with double- till sextuple-zeta quality basis sets have recently been reported in the literature. [193] Purpose of this Section is not to reproduce these results and describe accurately the dispersion forces between the noble gases, but rather to give a first glance on the performance of the approximate CCSD(T) scheme. For that reason, different approaches will be discussed, like which starting orbitals are optimal for the interference correction, and based on that, vital conclusions will be given. These conclusions will be used in the larger complexes described in the next Sections.

Figure 5.2 includes the potential energy curve of the helium dimer obtained from different levels of theory. As reference, the explicitly-correlated CCSD(T) method with the cc-pVQZ-F12 basis set was used and it is shown in the figure with the bold solid line. For the other methods under consideration, a triple-zeta quality basis was used (aug-cc-pVTZ). The choice of the specific basis set was done because it contains extra diffuse functions which are vital for the accurate description of noncovalent bonds. In addition, the CP correction was applied for the BSSE. The fixed F12 amplitudes approach was used for both CCSD(T)(F12) and interference-corrected MP2-F12 methods. The reason for not optimizing the F12 amplitudes is that the fixed-amplitudes approach has three advantages: firstly, its simplicity makes it favorable for large systems, secondly it is size-consistent, a vital virtue for studying weak interactions, and thirdly it removes the geminal BSSE. [194, 195] Finally, both canonical and localized orbitals have been used for the INT-MP2-F12. Results and differences between them are also one of the main topics of this Section.

All differences between the potential energy curves are extremely small (about 0.003 mE_h). However, the corresponding deviations from the reference curve lead to some interesting conclusions. Conventional CCSD(T) underestimates the weak interaction between the two helium atoms, while addition of the δE_{F12} term from MP2-F12 theory clearly overestimates it: the corresponding curve

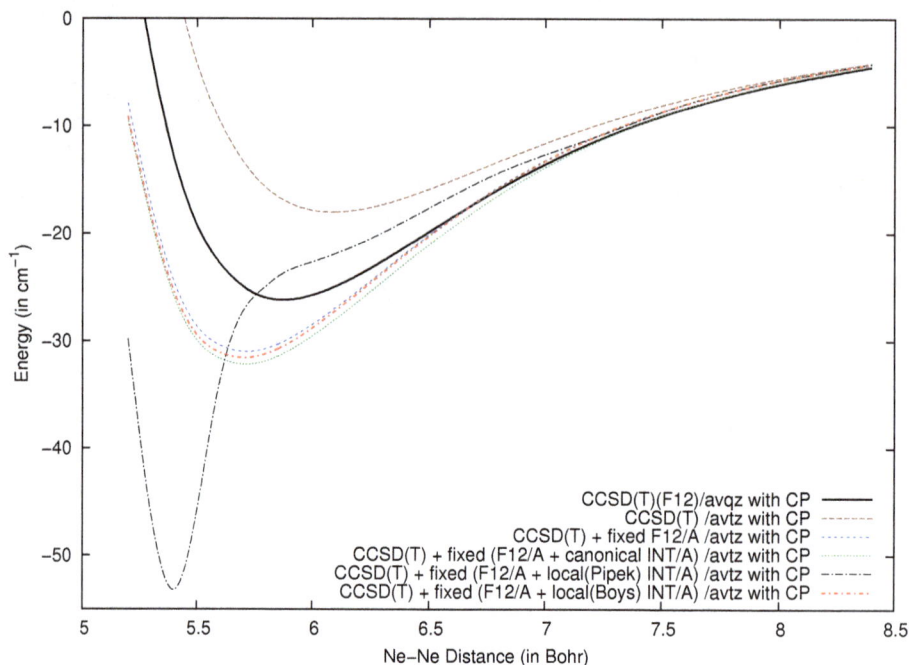

Figure 5.3: Potential energy curve of the neon dimer.

is significantly lower than the reference. Additionally, both curves deviate from the correct minimum, which is at ∼5.8 bohr. The CCSD(T) curve predicts a slightly longer interatomic distance while CCSD(T) with the F12 term slightly shorter. More accurate results, i.e. closer to the CCSD(T)(F12) curve, are obtained when the interference effects are taken into account, with or without applying a localization scheme. Boys and Pipek-Mezey localized orbitals for the helium dimer are identical, and both are closer to the reference values. At larger distances, all methods converge to the dissociation limit (not shown on the Figure).

Figure 5.3 shows the similar potential energy curves for the neon dimer. In this case too, conventional CCSD(T) underestimates the interaction energy between the two neon atoms, while, again, it predicts a minimum at a longer interatomic distance. Addition of second-order corrections overestimate the interaction energy at about $5\,\mathrm{cm}^{-1}$, in comparison to the CCSD(T)(F12) curve. On the contrary, the minimum of the potentials which include the second-order corrections are between -30.7 and -31.8 cm^{-1}, which are in a very good agreement with the CCSD(T)/aug-cc-pV5Z value (-28.1 cm^{-1}) [196] and the experimental value of -29.4 cm^{-1} [197]. At the same time, conventional CCSD(T) had a difference of about $10\,\mathrm{cm}^{-1}$.

The main issue which arises from interference effects in Figure 5.3 is the choice of orbitals for the calculation of the CCSD(T), INT and F12 terms. Both CCSD(T) and F12 energies are orbital invariant and thus, they are not affected by this choice. On the contrary, the INT-MP2-F12 theory is not, as it has been discussed in Chapter 3.2.2. The contribution of the δE_{INT} term calculated from canonical or localized orbitals with the Boys scheme is close to zero, but slightly negative. However, the differences between these three graphs (F12, F12 plus canonical-INT and F12 plus Boys-INT) are less than $1\,\mathrm{cm}^{-1}$. This means that the interference effects *do not contribute to*

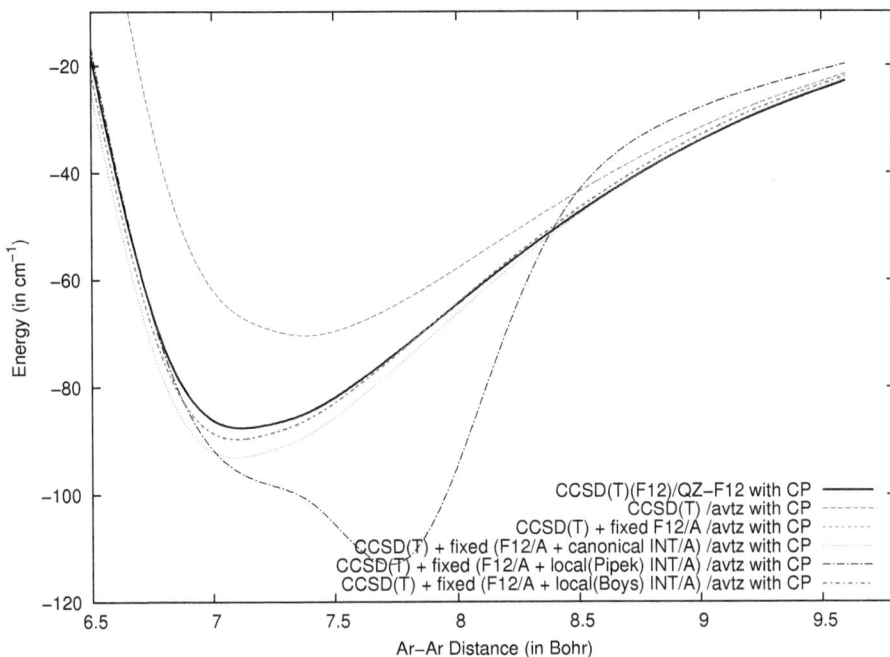

Figure 5.4: Potential energy curve of the argon dimer.

the dispersion, since the interference factors F^{ij} are about one for the interatomic pairs and the e_{ij}^{INT} pair energies almost zero. Indeed, for an interatomic distance of 5.84 α_0, the second-order basis-set truncation error from the interference-corrected MP2 (fixed amplitudes, approximation A, Boys LMOs) is 21.89 mE_h, exactly twice as the neon atom or the neon atom with the basis functions of the second neon atom (counterpoise correction). This behavior has also been pointed out for the most accurate of the CBS family of model chemistries, the CBS-QCI/APNO method. [7]

On the contrary, the description of this weakly interacting system deviates from reality significantly when the Pipek-Mezey localization scheme is applied. The calculated potential energy curve has no physical meaning, showing a very sharp minimum at -53 cm^{-1} at a shorter interatomic distance. The search for the source of this error should be focused on the resulting localized orbitals from the two schemes. Both schemes are diagonalizing the Fock matrix obtained from canonical orbitals, which are delocalized between the two neon atoms. The two procedures differ in how they construct the orbitals: the Pipek-Mezey scheme is based on the localization of the Mulliken atomic charge while Boys minimizes the sums of squared orbital radii (i.e. the spatial extension of the MOs by maximizing the square distance between charge centers). For the neon dimer, the Pipek-Mezey localization scheme produces four degenerate p-type LMOs, two of those localized at each neon atom. The other four valence LMOs are of sp-type, again localized at each neon atom. The energy gap between the first four LMOs and the rest is about half hartree. The sp-type LMOs split further in two pairs of degenerate LMOs, with a very small gap of 0.1 Hartree. On the contrary, eight sp^3-type valence LMOs result from the Boys localization scheme. Similar are also the orbitals from the monoatomic calculations where the basis set of the second neon is also present (formally is called "ghost" atom or monomer). However, the latter provides the correct potential energy curve, without producing artifacts.

The Boys localization scheme is also used in similar methods for intermolecular interaction energies, like the energy decomposition analysis of Azar and Head-Gordon. [108] In addition, Neese and co-workers [95] in their localized PNO methods and Tew and co-workers [94] in their local MP2-F12 theory with PNOs are using the Boys localization scheme, too. Therefore, the use of the Pipek-Mezey localized orbitals in the INT-MP2-F12 theory should be used with care.

It should also be noted that the early versions of the CBS model chemistries were using the Pipek-Mezey procedure, but this changed after the introduction of the CBS-QB3 model. [48] The original Pipek-Mezey scheme employs a unitary transformation of the occupied orbitals which maximizes the atomic populations based directly on the Mulliken analysis. A modification which was still maximizing the atomic populations of the transformed orbitals, but was measuring the populations in a minimum basis was introduced.

Increase of the interatomic distance leads to no interaction between the two neon atoms. CCSD(T)-INT-F12 calculations were carried out in an interatomic distance of 50 bohr and an energy of $6 \cdot 10^{-9}$ Hartree was obtained, independently of whether canonical or localized orbitals are used for the post-HF calculations (see also Section 3.3.6).

Similar trends have also been observed in the argon dimer potential energy curves (Figure 5.4). The second-order corrections of MP2-F12 or interference-corrected MP2-F12 are shifting the CCSD(T) curve close to the reference curve. Opposite to what was expected, the E_{INT} term from canonical orbitals is negative and moves further down the CCSD(T)-INT-F12 potential. However, the curve with the INT term from Boys LMOs is very close to the reference. The contribution of the specific term is very small and thus, the curve (with red color in Figure 5.4) coincides with the one which does not include interference effects. In the argon dimer too, the Pipek-Mezey procedure produces an erroneous potential energy curve, missing both the correct shape and the minimum of the potential.

In general, for many ground-state molecules, the orbitals localize clearly into atomic cores, bicentric bonds and lone pairs. Aromatic systems show larger delocalization, as expected and it will be discussed in the next Sections. In free radicals and similar systems, the unpaired electron is often delocalized, and this delocalization cannot be eliminated mathematically. Nevertheless, it is true even in radicals that most orbitals are well localized. [198]

5.2.3   Model acetonitrile dimer

The study of the noble gas dimers led to some first conclusions about the CCSD(T) with second-order energy terms from the interference-corrected MP2-F12. In particular, the applicability for studying noncovalent interactions with this method and its limitations were examined. However, some further analysis should be done, by focusing on some well defined systems which are dominated mainly from dispersion forces. Thus, a model geometry of the acetonitrile (CH_3CN) dimer (Figure 5.5) which was recently studied by Paul *et al.* [199] will be discussed in this Section. This choice has been done due to the small size of the model, while at the same time it includes some very interesting noncovalent phenomena.

Paul *et al.* [199] studied experimentally the charge density of the 2-methyl-4-nitro-1-phenyl-1H-imidazole-5-carbonitrile. The crystal packing is determined to some extend by the weak C-H$\cdots$O and C-H$\cdots$N hydrogen bonds but mostly by the lateral electrostatic interaction between the antiparallel C$\equiv$N groups. For an in-depth understanding of the interaction between the cyano groups, a smaller complex (shown in Figure 5.5) was isolated and studied with higher-level electronic structure methods. The 2-methyl-4-nitro-1-phenyl-1H-imidazole-part was substituted by a methyl-group, but the distance between the C≡N groups was kept the same as in the crystal structure.

Spin-component-scaled MP2-F12 (SCS-MP2-F12), conventional CCSD(T) and explicitly-correlated CCSD were used in that study for the calculation of the interaction energy of the model acetonitrile dimer. These methods estimate the noncovalent interaction at around 6 kcal/mol. Two DFT functionals were also tested (BP86 and B3LYP), but both catch only about 50% of the total interaction energy. On the contrary, when the empirical correction for van der Waals interactions of Grimme [159, 160] is added ("-D"), both functionals estimate the interaction energy close to the wave function methods, at -6.04 and -6.48 kcal/mol, respectively. All results, including DFT and CCSD(T) with second-order energy terms from interference-corrected MP2-F12 results, are shown in Table 5.1.

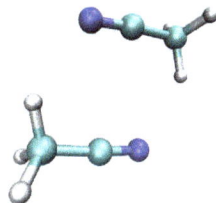

Figure 5.5: Model geometry of the acetonitrile dimer.

The experimental value for the dipole moment of the acetonitrile molecule is 3.9 D. This large value explains the electrostatic dipole-dipole nature of the dimer's interaction. Figure 11 of Ref. [199] shows clearly that the DFT functionals without the empirical van der Waals correction fail to include all the energy terms which describe the noncovalent interaction between the two acetonitrile molecules.

In order to obtain a more accurate estimate of the weak interaction of the specific complex, close to the basis set limit of the CCSD(T) method, explicitly-correlated CCSD(T) calculations with triple- and quadruple-zeta quality basis sets have been carried out. The counterpoise correction was taken into account in all calculations. Note that the deformation energy of the acetonitrile molecules was not added in the final interaction energy. The reason is that the acetonitrile dimer under study is just a model of one of the interactive sites of the larger 2-methyl-4-nitro-1-phenyl-1H-imidazole-5-carbonitrile crystal structure. In the original publication too, relaxation effects of the geometry had not been considered. Therefore, and in order to have a direct comparison with these results, the deformation energy of the monomers is not added.

Increase of the size of the basis set, from a double- to quadruple-zeta quality, leads to the convergence of the basis set limit of the noncovalent interaction between the model acetonitrile dimer. The interaction energy monotonically decreases as the basis set is increased (Table 5.1). However, the use of the explicitly-correlated coupled-cluster methods lead to a faster convergence: the CCSD(T)(F12) method with the double-zeta quality basis yields an energy almost identical with the one obtained from the conventional CCSD(T) with a basis set larger by one cardinal number (triple-zeta). From the results of the CCSD(T)(F12) with the larger basis sets (triple- and quadruple-zeta quality), an estimation for the interaction energy of the CH_3CN model dimer can be given. The energy obtained at the CCSD(T)(F12)/cc-pVTZ-F12 level of theory is -6.20 kcal/mol, more than 0.20 kcal/mol lower than the double-zeta result. Explicitly-correlated CCSD(T) with the cc-pVQZ-F12 basis set provides the best estimation for the interaction energy of this model, which is at about -6.3 kcal/mol.

The above interaction energies are compared with the corrected CCSD(T) energy from MP2-F12, by including interference effects. The cc-pVXZ-F12 basis sets were also used ($X = $ D,T,Q),

Table 5.1: Interaction energy of the model acetonitrile dimer at different levels of theory (in kcal/mol). The counterpoise correction was added for all methods.

Level of Theory	Interaction Energy	
	without "-D"[b]	with "-D"[b]
BP86/def2-TZVPP[a]	-3.02	-6.04
B3LYP/def2-TZVPP[a]	-3.46	-6.48
SCS-MP2-F12/cc-pVTZ-F12[a]	-5.68	
CCSD(F12)/cc-pVTZ-F12[a]	-5.71	
CCSD(T)/aug-cc-pVTZ[a]	-5.99	
CCSD(T)(F12)/cc-pVDZ-F12	-5.97	
CCSD(T)(F12)/cc-pVTZ-F12	-6.20	
CCSD(T)(F12)/cc-pVQZ-F12	-6.29	
	Canonical	Boys
CCSD(T)-INT-F12/cc-pVDZ-F12	-7.94	-6.30
CCSD(T)-INT-F12/cc-pVTZ-F12	-7.19	-6.32
CCSD(T)-INT-F12/cc-pVQZ-F12	-6.82	-6.35

[a] From Ref. [199].
[b] "-D" stands for the van der Waals correction of Grimme [160].

with the same auxiliary basis sets which were used in the calculations for the atomization energies and reaction barrier heights (Chapter 4). Two different types of orbitals for the post-HF calculations were used, canonical and localized from the Boys scheme. All results are shown in Table 5.1. Like in the CCSD(T)(F12) calculations, the CP correction was included, without adding the deformation energy.

CCSD(T)-INT-F12 with the Boys localization scheme yields interaction energies at the estimated basis set limit from explicitly-correlated CCSD(T). The calculated energies from the three different basis sets are in a range of only 0.05 kcal/mol. The CCSD(T) energy with the interference-corrected MP2-F12 terms obtained with the double-zeta quality basis is almost equal to the CCSD(T)(F12) energy from the quadruple-zeta basis. However, the main difference is the savings in the computational time gained from the new CCSD(T)-INT-F12 method: the CCSD(T)-INT-F12/cc-pVDZ-F12 calculation for the model dimer geometry of $(CH_3CN)_2$ needed 2 hours and 44 minutes while the CCSD(T)(F12)/cc-pVQZ-F12 calculation was running for more than a week (about 7 days and 6 hours). Both calculations have been carried out using only one processor core.

Even if the δE_{INT} contribution is relatively small, in comparison to the rest of the energy components, the addition of it is important for the accurate and quantitative description of the noncovalent interaction between the two CH_3CN molecules. The reason for its small magnitude is again the fact that the intermolecular interference factors are close to one and thus, the respective e_{ij}^{INT} pair energies close to zero.

On the contrary, interference-correction with canonical orbitals deviates significantly from the best estimate of the interaction energy of this model geometry. This observation puts their use for description of weakly interacting complexes in question. It has already been mentioned that the potential curves for the noble gas dimers obtained from CCSD(T)-INT-F12 with canonical orbitals are less accurate than the one with Boys LMOs. However, for the acetonitrile dimer, the difference of ca. 1.6 kcal/mol between the two different orbital schemes (cc-pVDZ-F12) is really large. Therefore, in the next Sections the Boys LMOs will be used, as far as the canonical orbitals should be avoided with the CCSD(T)-INT-F12 method for the examination of the noncovalent interactions.

One last comment should be given for the remaining methods which have been used by Paul *et al.*. The SCS-MP2-F12, CCSD(F12) and CCSD(T) energies are close to the basis set limit, but not as accurate as the CCSD(T)-INT-F12 method. The reason is that each of them lacks specific effects, which interference-corrected MP2-F12 achieves to include. For example, the CCSD(F12) method neglects the contribution of the perturbative triples; the CCSD(T) method misses the basis set limit which is gained from the explicitly-correlated methods. On the contrary, the scaling from the interference factors of the second-order energy term leads to a faster convergence to the basis set limit of the conventional CCSD(T) and yields very accurate interaction energies even with the use of a double-zeta quality basis.

The semi-empirical scaling of parallel and antiparallel spin pair correlation energies of the SCS-MP2-F12 method is a good approximation for shifting the MP2 basis set limit energies close to the QCISD(T) basis set limit [115]. However, it is a method which is not applicable to every system. A new parametrization [200] of the two factors (same-spin and opposite-spin), based on the S66 benchmark database [168] provided more accurate results on a test set of 38 data points which were not used for the parametrization. Different factors were obtained for different basis sets, and the corresponding RMS errors of the "SCS-S66-MP2" models were about half kcal/mol, while all the rest methods compared in that study yielded deviations of about 1 kcal/mol or more. Therefore, the SCS-MP2 coefficients can be described as "system dependent". This makes the specific method less suitable for noncovalent interactions, where a variety of different phenomena should be correctly described.

5.3   S22 Benchmark Database

5.3.1   Historical background

During the past years, Hobza and co-workers have published a series of articles with accurate hydrogen bonded and stacked DNA and RNA base pairs. [125,201,202] In their 2006 article [126], they reported the MP2 and CCSD(T) basis set limit interaction energies and geometries of 100 model complexes. Based on these results, they proposed a test set of 22 representative structures (S22) for noncovalent interactions in a balanced way. The geometries of all the members of the S22 test set are shown on Figure 5.6. It is important that the data set spans a wide range of interaction strengths to represent the diversity of interactions in biomacromolecules. [167] The benchmark database includes hydrogen bonded and electrostatically dominated complexes, complexes with predominant dispersion contribution and complexes with mixed noncovalent contributions. The database is balanced between these three categories and this helped to become a popular tool for designing or testing new faster computational methods, either for density functionals [161,192,203–207] or post-HF methods [97,139]. For example, Chai and Head-Gordon [205] optimized the long-range corrected density functional ωB97X-D based on a series of benchmark test sets, including molecules from G2, G3 and S22 databases, among others. However, the S22 data was weighted ten times more than the others.

In the original article of S22, the reference values of the interaction energies were estimated from the basis set limit of the CCSD(T) method, as obtained from Eq. (5.1). However, the complexity of the complexes included in the S22 database varies from very small dimers, like the water and methane dimers, to complexes with more than 15 non-hydrogen atoms, like the uracil dimer or

Hydrogen Bonded Complexes

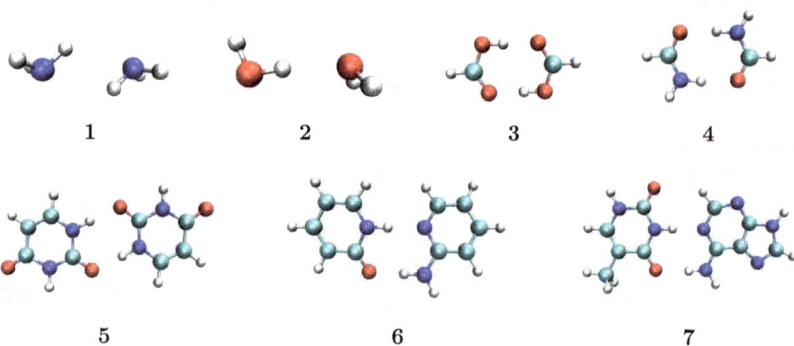

Dispersion Dominated Complexes

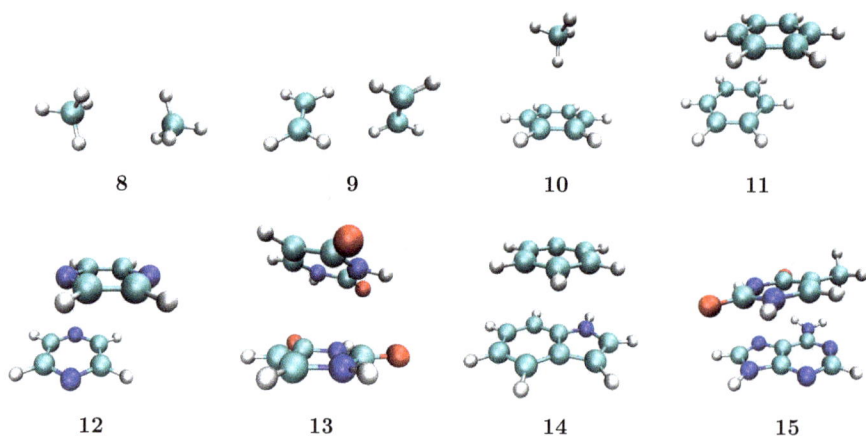

Mixed Complexes

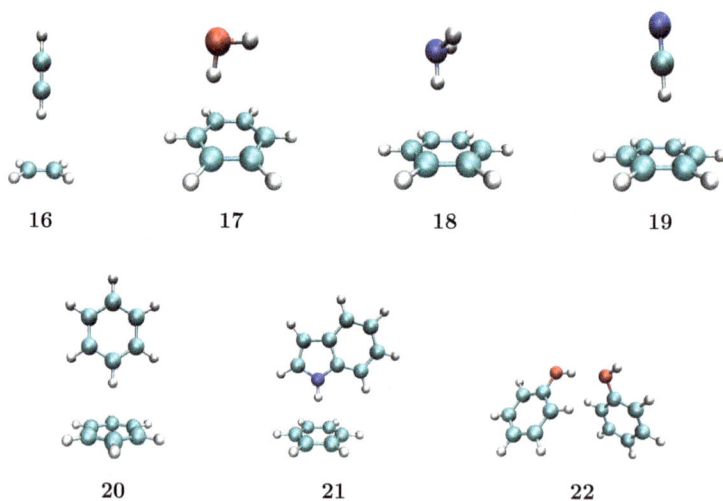

Figure 5.6: Equilibrium geometries of the S22 benchmark database (from Ref. [126]).

the adenine-thymine stacked geometry. For the latter, only small or medium size basis sets can be used for the coupled-cluster calculations. For that reason, different basis sets were used for these benchmark results, according to the size of each complex. Typically, for the small clusters, the CCSD(T) calculation was carried out with quadruple-zeta quality basis sets and the MP2/CBS was obtained from an quadruple/quintuple-zeta basis extrapolation (Eq. (2.9)). For the larger clusters, double- or triple-zeta basis sets were used for the coupled-cluster calculations.

A first revision of the benchmark interaction energies was published from Podeszwa *et al.* [208]. In their work, they recompute the CCSD(T)/CBS energies by using the same procedure as in the original work, i.e. by using Eq. (5.1), but with larger basis sets. In particular, they augment the basis sets with higher angular momentum, diffuse and mindbond functions. The corresponding average uncertainty was estimated to be about 1.0%.

The same year (2010) with the work of Podeszwa *et al.* [208] a similar revision of the S22 dataset was also published by Takatani *et al.* [209]. They also highlighted the insufficiency of some of the original S22 energies and improved the reference values by carrying out a series of calculations with different basis sets, but always based on Eq. (5.1). The new revision of interaction energies was designated S22A. Their most accurate model was based on the addition of the difference between extrapolated MP2/aug-cc-pV(DT)Z and CCSD(T)/aug-cc-pV(DT)Z correlation energies to the MP2/aug-cc-pV(TQ)Z energies. For the seven smallest complexes, the CCSD(T)/CBS limit was also estimated from a two-point extrapolation with the aug-cc-pVTZ and aug-cc-pVQZ basis sets. In addition, they computed these interaction energies with empirically corrected MP2 and CCSD methods. They concluded that Grimme's double-hybrid density-functional B2-PLYP [210] with an empirical correction for the dispersion forces (B2-PLYP-D) [204] outperforms all methods tested (MP2, SCS-MP2, SCS-CCSD, SCS(MI)-MP2 - the latter is an alternative parametrization of the SCS-MP2 for intermolecular interaction energies [211]). In particular, B2-PLYP-D has a MAD of only 0.12 kcal/mol, in respect to the S22A revision.

Liakos *et al.* [97] have also tested the accuracy of their parallel implementation of the local pair natural orbital coupled electron pair approach (LPNO-CEPA/1) in respect to the S22A revision. The total energy at the basis set limit was estimated from the sum of the HF energy, the MP2 correlation energy at the basis set limit and the higher-level correlation energy corrections, obtained from the difference of the LPNO-CEPA/1 and MP2 energies. For the HF energy, the aug-cc-pVQZ basis was used while for the higher-level term the aug-cc-pVTZ basis was used. The MP2 at the CBS limit was estimated from a two-point extrapolation by using Helgaker's formula (Eq. (2.10)). The extrapolation was carried out between the triple- and quadruple-zeta basis. In general, good accuracy with noteworthy speedup is achieved from this dual-basis scheme, where a MAD of 0.24 kcal/mol was reported. However, the maximum errors of 0.75, 0.76 and 0.79 kcal/mol for the uracil dimer (C_{2h}), the indole·benzene and the adenine·thymine complexes, respectively, imply that the LPNO-CEPA/1 should be used with caution.

One year later (2011) Marshall *et al.* [212] pointed out that the extrapolated energies from CCSD(T)/aug-cc-pV(TQ)Z are not always fully converged. Therefore, they provided a new revision of interaction energies named S22B. Table 5.2 includes the S22B reference values. Again, different levels of theory were used according to the size of the complexes. For the water dimer, the benchmark interaction energy was given as the CCSD(T) extrapolated energy from aug-cc-pV5Z and aug-cc-pV6Z basis sets, while for the rest six relatively small complexes the basis set limit was estimated from CCSD(T)/aug-cc-pV(Q5)Z. For the remaining fifteen complexes, the benchmark interaction energies were provided from Eq. (5.1), but with larger basis than in the original S22 and

S22A set of reference values. In particular, the basis set limit of the MP2 method was estimated from an aug-cc-pV(Q5)Z extrapolation, except for the T-shaped indole·benzene complex (Figure 5.6.**21**), which was estimated from an aug-cc-pV(TQ)Z extrapolation. The difference between the CCSD(T) and MP2 correlation energies was obtained from a triple-zeta quality basis set, except of the benzene·water (Figure 5.6.**17**), benzene·ammonia (Figure 5.6.**18**) and benzene·hydrogen cyanide complexes (Figure 5.6.**19**), which was obtained from the aug-cc-pVQZ basis set. It should be mentioned that not all energy components of Eq. (5.1) have been recomputed; most have been obtained from Ref. [208], [209] and [213].

The most accurate method to date, with respect to the S22B reference values, is the dispersion-weighted explicitly-correlated CCSD with the empirical (T**) scaling (DW-CCSD(T**)-F12). [140] The dispersion-weighted approach originates from the work of Marchetti and Werner. [139] They had proposed a weighted average between the conventional and the SCS-MP2 methods, named as DW-MP2. The system dependent factor w is the switching function

$$w = \frac{1}{2}\left[1 + \tanh\left(a + b\frac{\Delta E_{\text{SCF}}}{\Delta E_{\text{MP2-F12}}}\right)\right], \tag{5.6}$$

where a and b are fitting parameters and they have been determined by a least-squares fit of the MP2-F12 energies to the CCSD(T*)-F12a results for the S22 test set. The ratio between the HF and MP2-F12 energies serves as an inspector for the nature of the interactions dominating the complex under study: for dispersion-dominated systems, the ratio is negative, while for electrostatic-dominated complexes, the ratio is positive. Despite the empiricism behind this method, DW-MP2 achieves a RMS deviation of 0.24 kcal/mol for the S22B revision. Based on this idea, a similarly weighted sum between the CCSD(T**)-F12a and the CCSD(T**)-F12b interaction energies was proposed. [140] The a and b parameters were fitted against the S22B test set, and set as -1 and 4, respectively. The DW-CCSD(T**)-F12 method has a MAD of 0.05 kcal/mol with a 0.07 kcal/mol RMS deviation, while the CCSD(T**)-F12a and CCSD(T**)-F12b interaction energies have a MAD of 0.12 and 0.10 kcal/mol, respectively.

5.3.2 CCSD(T)-INT-F12 against S22B

The performance of the CCSD(T) method with second-order terms from interference-corrected explicitly-correlated MP2 on the interaction energies of the S22 database was examined. The S22B revision was used as reference for the 22 noncovalent bonded structures as far as it is the latest and most accurate from the original S22 and the S22A revision, for reasons explained above.

The fixed-amplitude 2A ansatz with the F + K approximation [102] for the commutator of kinetic energy with Slater-type geminal f_{12} was used for the interference-corrected explicitly-correlated MP2 calculations. The advantages of the fixed-amplitude approach have been mentioned in the previous Section. The exponents of the f_{12} function were taken as recommended by Peterson and co-workers for the two basis sets used (0.9 for cc-pVDZ-F12 and 1.0 for cc-pVTZ-F12). The cc-pVXZ-F12 auxiliary basis for CABS was used, where X is the cardinal number of the basis set. The aug-cc-pwCV$(X + 1)Z$ cbas of Hättig [120] (aug-cc-pV$(X + 1)Z$ for the hydrogens) was used for the robust fitting of both the F12 integrals and the usual electron-repulsion integrals. The aug-cc-pV$(X + 1)Z$ jkbas basis of Weigend [121] was used for the two-electron contributions to the Fock matrix. The active occupied orbitals were localized with the Boys localization scheme. [105, 106]

The original S22 optimized geometries of Jurecka *et al.* [126] were used in order to have a direct comparison with the reference values and literature data from other electronic structure theory

Table 5.2: S22B benchmark values and interaction energies for the S22 database (in kcal/mol) calculated from conventional CCSD(T), CCSD(T) without (CCSD(T)-F12) and with interference effects (CCSD(T)-INT-F12). The cc-pVDZ-F12 was used for all methods. In parentheses are the differences from the reference energies.

Complex	S22B[a]	CCSD(T)		CCSD(T)-F12[b]		CCSD(T)-INT-F12	
Hydrogen Bonded Complexes							
1 $(NH_3)_2$, C_{2h}	-3.133	-2.726	(-0.41)	-3.136	(0.00)	-3.121	(-0.01)
2 $(H_2O)_2$, C_s	-4.989	-4.423	(-0.57)	-4.979	(-0.01)	-4.995	(0.01)
3 Formic Acid Dimer, C_{2h}	-18.753	-16.321	(-2.43)	-18.694	(-0.06)	-18.833	(0.08)
4 Formamide Dimer, C_{2h}	-16.062	-14.227	(-1.84)	-15.987	(-0.08)	-16.025	(-0.04)
5 Uracil Dimer, C_{2h}	-20.641	-18.725	(-1.92)	-20.544	(-0.10)	-20.633	(-0.01)
6 2-pyridoxine-2-aminopyridine, C_1	-16.934	-15.119	(-1.81)	-16.832	(-0.10)	-16.997	(0.06)
7 Adenine-Thymine WC, C_1	-16.660	-14.784	(-1.88)	-16.554	(-0.11)	-16.553	(-0.11)
Dispersion Dominated Complexes							
8 $(CH_4)_2$, D_{3d}	-0.527	-0.373	(-0.15)	-0.495	(-0.03)	-0.498	(-0.03)
9 $(C_2H_4)_2$, D_{2d}	-1.472	-1.045	(-0.43)	-1.450	(-0.02)	-1.453	(-0.02)
10 Benzene-CH$_4$, C_3	-1.448	-1.082	(-0.37)	-1.408	(-0.04)	-1.411	(-0.04)
11 Benzene Dimer Stack, C_{2h}	-2.654	-1.723	(-0.93)	-2.555	(-0.10)	-2.595	(-0.06)
12 Pyrazine Dimer, C_s	-4.255	-3.038	(-1.22)	-4.149	(-0.11)	-4.201	(-0.05)
13 Uracil Dimer, C_2	-9.805	-8.068	(-1.74)	-9.615	(-0.19)	-9.687	(-0.12)
14 Indole-Benzene, C_1	-4.524	-3.234	(-1.29)	-4.336	(-0.19)	-4.341	(-0.18)
15 Adenine-Thymine Stack, C_1	-11.730	-9.560	(-2.17)	-11.523	(-0.21)	-11.599	(-0.13)
Mixed Complexes							
16 Ethene-Ethine, C_{2v}	-1.496	-1.235	(-0.26)	-1.467	(-0.03)	-1.460	(-0.04)
17 Benzene-H$_2$O, C_s	-3.275	-2.750	(-0.52)	-3.215	(-0.06)	-3.208	(-0.07)
18 Benzene-NH$_3$, C_s	-2.312	-1.881	(-0.43)	-2.267	(-0.05)	-2.266	(-0.05)
19 Benzene-HCN, C_s	-4.541	-3.846	(-0.70)	-4.493	(-0.05)	-4.463	(-0.08)
20 Benzene Dimer T-shaped, C_1	-2.717	-2.126	(-0.59)	-2.657	(-0.06)	-2.660	(-0.06)
21 Indole-Benzene T-shaped, C_1	-5.627	-4.822	(-0.81)	-5.571	(-0.06)	-5.548	(-0.08)
22 Phenol Dimer, C_1	-7.097	-5.924	(-1.17)	-6.933	(-0.16)	-7.076	(-0.02)
Mean Error		-1.07		-0.08		-0.05	
MAD		1.07		0.08		0.06	
RMS		1.27		0.10		0.07	
Max Error		-2.43		-0.21		-0.18	

[a] From Ref. [212]
[b] Second-order F12 energy term.

Table 5.3: CCSD(T)-INT-F12/cc-pVTZ-F12 interaction energies (in kcal/mol) for selected complexes of the S22 benchmark database. In parentheses are the differences from the S22B[a] reference energies.

	Complex	CCSD(T)-INT-F12	
1	$(NH_3)_2$, C_{2h}	-3.129	(0.00)
2	$(H_2O)_2$, C_s	-4.988	(0.00)
3	Formic Acid Dimer, C_{2h}	-18.804	(0.05)
4	Formamide Dimer, C_{2h}	-16.062	(0.00)
8	$(CH_4)_2$, D_{3d}	-0.511	(-0.02)
9	$(C_2H_4)_2$, D_{2d}	-1.475	(0.00)
10	Benzene·CH$_4$, C_3	-1.433	(-0.02)
16	Ethene·Ethine, C_{2v}	-1.491	(-0.01)
17	Benzene·H$_2$O, C_s	-3.242	(-0.03)
18	Benzene·NH$_3$, C_s	-2.303	(-0.01)
19	Benzene·HCN, C_s	-4.518	(-0.02)
	Mean	0.00	
	MAD	0.01	
	RMS	0.02	
	Max Error	0.05	

[a] From Ref. [212]

methods. All energies calculated with the CCSD(T)-INT-F12 theory include a series of different energy terms:

$$E_{CCSD(T)/CBS} \approx E_{CCSD(T)} + \delta E_{INT} + \delta E_{F12} + \delta E_{CABS}. \tag{5.7}$$

These are the Hartree-Fock energy, the conventional CCSD(T) correlation energy, the second-order terms of MP2-F12 scaled for every electron pair with the interference factors ($\delta E_{INT/A}$ and $\delta E_{F12/A}$), and the Hartree-Fock correction from the F12 theory (CABS singles).

The interaction energies obtained at the CCSD(T)-INT-F12/cc-pVDZ-F12 level are compared with the S22B reference values, the conventional coupled-cluster and the CCSD(T) energies corrected from the $\delta E_{F12/A}$ term of the second-order MP2-F12 theory. All results and reference values are shown on Table 5.2, with the corresponding statistical deviations (mean error, mean absolute deviation, root-mean-square error and maximum error). The "kcal per mol" units have been preferred in this Section in order to be consistent with the literature and having a direct comparison with statistical deviations from other methods. All interaction energies include the counterpoise correction (Eq. (5.2)) for the treatment of the BSSE error.

As it has already been explained thoroughly in this Thesis, conventional coupled-cluster methods need large basis sets in order to reach the correlation energy basis set limit. This observation also holds for the interaction energies of the 22 complexes. The deviations from the S22B reference data vary from 0.15 to almost 2.5 kcal/mol. These errors are typically smaller for the small complexes, like the water dimer (02) or the methane dimer (08), and they increase with increasing size. The only exception is the formic acid dimer (03), a relatively small complex which consists of only six non-hydrogen atoms. The deviation from the CCSD(T)/cc-pVDZ-F12 level of theory is significantly large (-2.43 kcal/mol) which is the maximum error from these 22 structures. The large distribution of deviations is reflected by the large RMS value of 1.27 kcal/mol (5.31 kJ/mol), where the MAD exceeds 1 kcal/mol. Therefore, the errors introduced from the small basis set are too large for a quantitative description of the noncovalent interactions of the 22 complexes.

Addition of the second-order explicitly-correlated energy term δE_{F12} alleviates the intrinsic deficiency of the CCSD(T) method with a small basis. The differences from the S22B revision reference data are about ten times smaller. The MAD and RMS values of 0.08 kcal/mol (0.33 kJ/mol) and 0.10 kcal/mol (0.42 kJ/mol), respectively, show clearly the significant improvement on the description of the noncovalent interactions for all the 22 complexes. The maximum error was found for the stacked adenine·thymine complex (-0.21 kcal/mol). For all the complexes, the deviations are associated with the number of electron pairs and increase with increasing number of pairs. However, there is one notable exception. Although the same number of electron pairs (4753) for the two adenine·thymine complexes (No. 7 and 15), the electrostatic (hydrogen) bonded complex has almost a difference from the reference values two times smaller (-0.11 for No. 7, -0.21 for No. 15). In combination with the deviations found for the stacked conformations of the uracil dimer and the indole·benzene complex, a conclusion can be drawn: small deviations for the dispersion dominated complexes should be expected. It should also be mentioned that, like in the conventional CCSD(T), the addition of the F12 energy component constantly underestimates the interaction energies of the 22 complexes. This is shown from the negative deviations and makes all description of noncovalent interactions with this method more consistent.

Smaller deviations are observed when interference effects are included in the description of the noncovalent interactions. Even if the differences from the CCSD(T) with the second-order F12 term are negligible for the small members of the test set, better accuracy is achieved for the larger complexes. This leads to a further decrease of the statistical deviations: the MAD of 0.06 kcal/mol (0.25 kJ/mol) and the RMS of 0.07 kcal/mol (0.29 kJ/mol) are an improvement over the two previous methods discussed above. In addition, this level of accuracy is the highest reported in literature, together with the DW-CCSD(T**) of Marshall *et al.* [140]. However, the CCSD(T)-INT-F12 includes no empirical parameters nor scalings, increasing the applicability of the method to more applications, like thermochemistry, as it has been shown in the previous Chapter.

All results showed in Table 5.2 have been obtained with a double-zeta quality basis (cc-pVDZ-F12). This highlights one more advantage of the interference-effects. This level of accuracy is achieved with a relative low computational effort. The most demanding calculation was the stacked adenine·thymine complex (No. 15). The structure of the dimer is composed of 19 non-hydrogen atoms and the corresponding calculation needed 14 days and 12 hours in eight cores of one Intel Xeon X5460 processor, with 32GB of memory.

Improved accuracy can be achieved when a triple-zeta quality basis is used. The interaction energies of 11 out of 22 complexes have been calculated with the cc-pVTZ-F12 basis set and the CCSD(T)-INT-F12 results are shown in Table 5.3. This is verified from the very low RMS value of 0.02 kcal/mol (0.08 kJ/mol). Unfortunately, calculations for the complexes with more than 12 non hydrogen atoms require too much computational effort (in terms of memory and disk space) and thus, they have been excluded from the this study.

5.4 Imidazole·Benzene and Pyrrole·Benzene Complexes

Interactions between aromatic rings are central to many areas of modern chemistry; they are very frequent in biomolecules, nanomaterials and organic/inorganic crystal structures. [149, 167, 214]

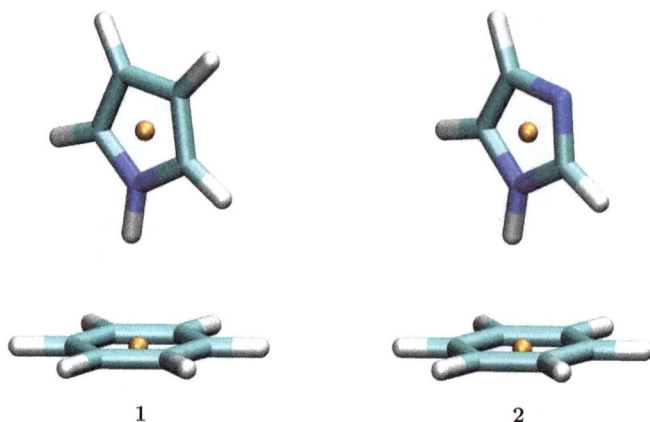

Figure 5.7: Ground state geometries of (1) pyrrole·benzene and (2) imidazole·benzene complexes. The golden spheres represent the center of mass of each aromatic molecule.

The driving forces of stabilization of these type of noncovalently bonded systems are dispersion and/or induction forces. These forces lead to a competition of different possible ground state geometries for this kind of complexes, like π-π stacked, T-shaped or planar conformations. Further understanding of the nature of interaction and the preferable stable geometries of such structures can be provided by studying model noncovalent dimers. Such models have already been discussed in Section 5.3; the S22 benchmark database includes some typical examples, like the stacked adenine·thymine complex or the two different conformations of the benzene dimer (π-π stacked and T-shaped geometries).

Recently, the imidazole·imidazole, benzene·imidazole and benzene·indole dimers were studied. [215] Their best estimate for the binding energies of these complexes were based on the CBS limit of CCSD(T), as it has been obtained from Eq. (5.1). The MP2 energy at the basis set limit was estimated from a two-point extrapolation from the energies calculated with the aug-cc-pVDZ and aug-cc-pVTZ basis sets. The authors have concluded that the imidazole·imidazole dimer has a preference for a hydrogen bonded geometry between the two imidazole rings. On the contrary, a T-shaped geometry is more likely for the benzene·imidazole and benzene·indole dimers. The above are based on the calculated binding energies and the structural parameters. Symmetry-adapted perturbation theory (SAPT) calculations provided more evidence for these findings.

In this Section, two structures with two aromatic molecules are investigated. The first is the imidazole·benzene complex, which was also present in the study of Karthikeyan and Nagase [215]. The second is the pyrrole·benzene complex which was studied both experimentally and computationally from Pfaffen et al. [216]. Both complexes can be considered as models for simulate weak interactions between the side chains of specific amino-acids. The pyrrole moiety serves as a mimic for the side chain of proline, the imidazole moiety for histidine and the benzene ring as a mimic for the side chain of phenylalanine or tyrosine.

These two complexes are examined together because they share the same non-conventional hydrogen bond between a NH donor and the π-electron cloud of the benzene moiety. These T-shaped amino NH···π aromatic interactions are important determinants of secondary structures of peptides and proteins. Typical dimers which exhibit such NH$\cdots\pi$ hydrogen bonds are the ammonia·benzene complex [217–219] (which is included in the S22 database too, Figure 5.6.**18**), the pyrrole dimer [220–222] and the 2-pyridone·benzene complex [223, 224]. Further quantitative description of these noncovalent bonds can be provided from the two complexes studied in this Section.

The ground state energy minimum structures were optimized at the SCS-MP2/aug-cc-pVTZ level of theory by Chantal Pfaffen of Univerität Bern. Both structures have a C_s symmetry and are shown in Figure 5.7. The golden spheres represent the center of mass of each aromatic molecule. For the pyrrole·benzene complex (Figure 5.7.1), the distance between the two centers is 8.22 a_0 and for the imidazole·benzene complex (Figure 5.7.2) this distance is at 7.86 a_0. The distance of the nearest H atom from the benzene plane is 4.24 a_0 for the pyrrole· and 4.22 a_0 for the imidazole·benzene complex.

The interaction energy between the monomers of these two complexes was computed either without (w/o CP; ΔE_2) or with applying a counterpoise correction (w/ CP; ΔE_2^{cp}). These two-body terms have been described in Section 5.1.1 (Eqs. (5.3)). Different levels of theory were used for the calculation of the interaction energies. These can be separated in two groups. The first includes conventional methods (i.e. non explicitly-correlated), like the MP2, CCSD and CCSD(T) methods, plus two empirically scaled MP2 variants, the SCS-MP2 [115] and the DW-MP2 [139]. The second group includes the explicitly-correlated variants of these methods. All results obtained from these methods are shown in Tables 5.4 and 5.5. Three different basis sets were used, the aug-cc-pVXZ basis, where X corresponds to D, T and Q quality.

The best estimations for the interaction energies of the imidazole·benzene and pyrrole·benzene complexes have been acquired from the basis set limit of the CCSD method, as it has been calculated from the explicitly-correlated CCSD method with the aug-cc-pVQZ basis set. On top of these interaction energies, the empirical scaled perturbative triples (T*) are added. However, due to the large computational effort needed to calculate the (T) energy with a quadruple-zeta quality basis, this energy component has been obtained with the smaller aug-cc-pVTZ basis. Therefore, the necessity for the empirical scaling is important, in order to acquire a more accurate (T) contribution, close to the basis set limit. For the first complex, the best estimation for the ΔE_2^{cp} is -22.74 kJ/mol. The interaction energy of the pyrrole·benzene is about 2 kJ/mol higher, at -20.38 kJ/mol. Based on these values, a discussion on the results obtained from the different levels of theory will follow.

As it is well known, Hartree-Fock theory fails on the qualitative and quantitative description of non-covalent bonds. Therefore, the need for accurate post-Hartree-Fock methods is more than obvious. The lowest level of theory which includes correlation energy terms is perturbation theory. MP2 theory [70] has been proven a useful tool for calculating relative energies, mainly due to its low computational cost, in comparison with more sophisticated methods, like coupled-cluster theory. However, it usually overestimates interaction energies at its basis set limit and thus, it should be used with caution. For example, the ΔE_2^{cp}, as it has been obtained from the MP2/aug-cc-pVDZ level of theory for the imidazole·benzene complex (Table 5.4), is -24.01 kJ/mol. With the augmented triple-zeta quality basis set, the interaction energy is not increased in order to move closer to the best estimate, but it is reduced further to -26.61 kJ/mol. The same holds for the explicitly-correlated variant of MP2, which leads to a faster convergence to the basis set limit of this method. In other words, it leads to a "faster overestimation" of the interaction energy. The counterpoise-corrected MP2-F12 values for the imidazole·benzene complex are decreased

Table 5.4: Interaction energy (ΔE_2 in kJ/mol) of the imidazole·benzene complex at its SCS-MP2/aug-cc-pVTZ geometry.

Method	aug-cc-pVDZ		aug-cc-pVTZ		aug-cc-pVQZ[a]	
	w/o CP	w/ CP	w/o CP	w/ CP	w/o CP	w/ CP
HF	-6.80	-1.99	-3.26	-2.14	-	-
MP2	-38.75	-24.01	-33.23	-26.61	-	-
SCS-MP2	-33.13	-18.22	-27.72	-20.57	-	-
DW-MP2	-37.21	-23.79	-31.43	-24.88	-	-
CCSD	-29.91	-16.14	-24.27	-18.32	-	-
CCSD(T)	-34.43	-19.12	-28.13	-21.74	-	-
HF+CABS	-3.29	-2.04	-2.34	-2.11	-2.20	-2.11
MP2-F12	-29.72	-26.94	-28.67	-27.66	-28.19	-27.72
SCS-MP2-F12	-23.69	-21.10	-22.68	-21.68	-22.21	-21.74
DW-MP2-F12	-27.75	-25.67	-26.57	-25.66	-26.08	-25.64
CCSD(F12)	-21.25	-18.47	-20.07	-19.00	-19.42	-19.05
CCSD(F12)+(T)	-25.78	-21.45	-23.93	-22.43	-23.28	-22.48
CCSD(F12)+(T*)	-26.36	-22.14	-23.93	-22.68	-23.27	-22.74

[a] Triples corrections (T) and (T*) computed in the aug-cc-pVTZ basis.

Table 5.5: Interaction energy (ΔE_2 in kJ/mol) of the pyrrole·benzene complex at its SCS-MP2/aug-cc-pVTZ geometry.

Method	aug-cc-pVDZ		aug-cc-pVTZ		aug-cc-pVQZ[a]	
	w/o CP	w/ CP	w/o CP	w/ CP	w/o CP	w/ CP
HF	-4.16	0.82	-0.47	0.64	-	-
MP2	-36.99	-21.95	-31.17	-24.42	-	-
SCS-MP2	-31.31	-16.11	-25.52	-18.33	-	-
DW-MP2	-35.15	-22.22	-28.85	-22.62	-	-
CCSD	-27.92	-13.89	-21.89	-15.91	-	-
CCSD(T)	-32.58	-17.00	-25.88	-19.47	-	-
HF+CABS	-0.56	0.73	0.44	0.67	0.58	0.67
MP2-F12	-27.55	-24.67	-26.40	-25.37	-25.90	-25.43
SCS-MP2-F12	-21.44	-18.77	-20.34	-19.33	-23.86	-23.38
DW-MP2-F12	-25.07	-23.15	-23.73	-22.89	-24.99	-24.61
CCSD(F12)	-18.87	-16.01	-17.59	-16.51	-16.92	-16.56
CCSD(F12)+(T)	-23.52	-19.12	-21.58	-20.06	-20.91	-20.11
CCSD(F12)+(T*)	-24.11	-19.81	-21.58	-20.32	-20.90	-20.38

[a] Triples corrections (T) and (T*) computed in the aug-cc-pVTZ basis.

from -26.94 to -27.72 kJ/mol as the basis set is increased. Same behavior is observed also for the pyrrole·benzene complex; the MP2 interaction energy at the basis set limit is about -25.4 kJ/mol.

Alternative methods which can in principle alleviate the overestimation of the MP2 basis set limit interaction energies, but simultaneously keeping the same computational effort of MP2 have become very popular. Examples of such (empirical) methods are the SCS-MP2 and the DW-MP2 which have been analyzed in the previous Sections. Both have been parametrized in order to provide (with a small basis set) high accuracy, typically obtained at the CCSD(T) level with a large basis set. However, SCS-MP2 clearly underestimates the interaction energies for both complexes.

Same holds also for the basis set limit of the method, as it has been calculated from the F12 variant (SCS-MP2-F12). On the contrary, DW-MP2 overestimates the strength of the interaction. In particular, and quite surprisingly, the interaction energies are becoming lower by increasing the basis set. This means that by increasing basis size, the deviation gets larger. At the same time, the DW-MP2-F12 values for the imidazole·benzene complex are almost not affected by the size of the basis set used (ca. -25.65 kJ/mol) while oscillations are observed for the pyrrole·benzene complex. All DW-MP2(-F12) interaction energies are far from the best estimate. Therefore, these methods should be used with caution for non-trivial cases.

The CCSD level of theory which neglects any contribution from triples excitations provides very poor interaction energies. For both complexes, the CCSD/aug-cc-pVTZ energies have a deviation of about 4 kJ/mol from the corresponding best estimations. CCSD(F12) leads to a faster convergence to the basis set limit of the method: about -19.1 kJ/mol for imidazole·benzene and -16.6 kJ/mol for pyrrole·benzene. However, the need for inclusion of the pertubative triples is more than necessary for a quantitative description of the amino NH···π aromatic interaction.

The slow convergence of the conventional CCSD(T) and the essence of the explicitly-correlated variant is demonstrated nicely in Tables 5.4 and 5.5. For both systems, the CCSD(T) interaction energies in the aug-cc-pVTZ basis set are about one kJ/mol higher than the best estimate. On the contrary, CCSD(F12) with the addition of the perturbative triples leads to faster convergence to the basis set limit, even with the use of a triple-zeta quality basis set. In particular, the addition of the empirical (T*) term increases the accuracy of the method for both complexes. For example, the interaction energy of the imidazole·benzene complex at the CCSD(T)(F12)/aug-cc-pVDZ level of theory without the (T*) contribution is -21.45 kJ/mol, 1.3 kJ/mol higher than the best estimate. This energy difference drops to 0.6 kJ/mol with the addition of the (T*) correction.

In principle, in a formally complete basis set, the BSSE is eliminated. This means that (often but not always) the larger the basis set, the smaller the counterpoise correction is. Indeed, comparison between the interaction energies of all methods (Tables 5.4 and 5.5) show that the contribution of the counterpoise correction, i.e. the difference between ΔE_2^{cp} and ΔE_2, is smaller when the basis set limit is approached. This is done either when the basis set is increased or when explicitly-correlated theory is used. For example, for the pyrrole·benzene complex at the CCSD(T)/aug-cc-pVDZ level of theory, the contribution of the counterpoise correction is 15.58 kJ/mol. At the same time, when the explicitly-correlated variant is used, the contribution drops to 7.40 kJ/mol. However, the correction still remains significantly high and its neglection should be avoided.

The performance of the method presented in this Thesis was also tested for the calculation of the interaction energies of the imidazole·benzene and pyrrole·benzene complexes. [225] All the individual terms which are summed in the CCSD(T)-INT-F12 method are shown in Table 5.6. Double- and triple-zeta quality basis sets were tested. The second-order corrections from the interference-corrected MP2-F12 scheme were calculated with either the aug-cc-pVXZ basis sets or the cc-pVXZ-F12 basis, where X stands for D or T, while the CCSD(T) interaction energies have been obtained with the aug-cc-pVXZ family of basis sets (Tables 5.4 and 5.5). The CABS singles correction was also added. This contribution was obtained from the same basis sets which were used for the CCSD(T) calculations. The reason for this choice is that this specific energy component constitutes a correction to the Hartree-Fock energy, which is included in the CCSD(T) energies. Results with and without counterpoise correction are also included in Table 5.6.

Table 5.6: Interaction energy (ΔE_2 in kJ/mol) calculated from CCSD(T) with second-order terms from interference-corrected MP2-F12.

		Imidazole·Benzene		Pyrrole·Benzene	
		w/o CP	w/ CP	w/o CP	w/ CP
CCSD(T)	aVDZ	-34.43	-19.12	-32.58	-17.00
CABS	aVDZ	3.51	-0.05	3.60	-0.09
F12	aVDZ	6.60	-3.24	6.95	-2.97
INT	aVDZ	-0.84	-0.19	-0.88	-0.18
CCSD(T)-INT-F12		-25.16	**-22.60**	-22.91	**-20.24**
CCSD(T)	aVDZ	-34.43	-19.12	-32.58	-17.00
CABS	aVDZ	3.51	-0.05	3.60	-0.09
F12	DZ-F12	1.23	-3.67	1.60	-3.40
INT	DZ-F12	-0.20	0.09	-0.29	0.07
CCSD(T)-INT-F12		-29.90	**-22.75**	-27.67	**-20.42**
CCSD(T)	aVTZ	-28.13	-21.74	-25.88	-19.47
CABS	aVTZ	0.92	0.03	0.91	0.03
F12	aVTZ	4.21	-1.17	4.35	-1.07
INT	aVTZ	-1.34	0.00	-1.37	-0.01
CCSD(T)-INT-F12		-24.33	**-22.88**	-21.99	**-20.52**
CCSD(T)	aVTZ	-28.13	-21.74	-25.88	-19.47
CABS	aVTZ	0.92	0.03	0.91	0.03
F12	TZ-F12	1.03	-1.19	1.13	-1.10
INT	TZ-F12	-0.35	0.02	-0.38	0.01
CCSD(T)-INT-F12		-26.53	**-22.88**	-24.22	**-20.53**

[a] aVX is an abbreviation for the aug-cc-pVXZ family of basis sets; XZ-F12 for the cc-pVXZ-F12 family.

As it has already been mentioned, conventional CCSD(T) with either double- or triple-zeta basis sets does not provide a quantitative result for the strength of the noncovalent interactions. This drawback is alleviated from the contribution of the second-order δE_{F12} term, in conjunction with interference-effects. The F12 terms calculated with the aug-cc-pVDZ basis set (for both complexes) are not adequate to reproduce the best estimation for the interaction energies. On the contrary, the addition of the δE_{INT} term shifts the interaction energy very close to the CCSD(F12)+(T*) estimate; the counterpoise corrected result is 0.14 kJ/mol higher than the best estimate, for both imidazole·benzene and pyrrole·benzene complexes.

Excellent agreement with the CCSD(F12)+(T*) interaction energy is achieved when the correction terms from INT-MP2-F12/cc-pVDZ-F12 are added to the CCSD(T)/aug-cc-pVDZ energies. The corresponding results for imidazole·benzene (-22.75 kJ/mol) and pyrrole·benzene (-20.42 kJ/mol) differ only by 0.01 and 0.04 kJ/mol (in absolute values) from the best estimations (shown with bold font in Table 5.6). This observation suggests that the interference-corrected explicitly-correlated MP2 in conjunction with the cc-pVDZ-F12 basis set can yield very accurate results, like it has been shown for the S22 benchmark database (Section 5.3).

The main difference between the correcting terms obtained with aug-cc-pVDZ and cc-pVDZ-F12 basis is that the $\delta E_{\text{F12/cc-pVDZ-F12}}$ term overestimates (i.e. yields larger energies, in absolute values) the interaction energy of both complexes. Therefore, the $\delta E_{\text{INT/cc-pVDZ-F12}}$ term is positive and corrects this trend. The opposite happened for the terms calculated with the aug-cc-pVDZ basis. The δE_{F12} term is smaller than the one from the cc-pVDZ-F12 basis. This discrepancy is corrected by the interference effects: the $\delta E_{\text{INT/aug-cc-pVDZ}}$ term is negative and thus, it shifts the total CCSD(T)-INT-F12 interaction energy to the best estimation of the basis set limit of coupled-cluster.

The magnitude of the correcting terms calculated with the triple-zeta basis sets are significantly smaller, as expected. CCSD(T) with triple-zeta quality basis achieves to gain more correlation energy and yields results closer to the basis set limit. However, complementary terms which will provide higher accuracy are still needed. The δE_{F12} terms are about one third of those from the double-zeta results and they are adequate to recover the missing correlation energy. Therefore, interference effects are very small, close to zero kJ/mol, for these two compounds at the triple-zeta level.

The total CCSD(T)-INT-F12 interaction energies calculated either with the aug-cc-pVTZ or cc-pVTZ-F12 basis sets are identical; -22.88 kJ/mol for the imidazole·benzene complex and -20.52 kJ/mol for the pyrrole·benzene, as they are shown in Table 5.6 with bold font. Both values are 0.14 kJ/mol lower than the CCSD(F12)+(T*) interaction energies. This difference generates automatically one important question: which is the correct basis set limit? Both methods suppose to give the same result, but using different theoretical assumptions. Based on the analysis of the previous Sections, CCSD(T)-INT-F12 with a triple-zeta basis yields interaction energies at the CCSD(T) basis set limit. CCSD(F12) with a triple-zeta basis set can adequately reproduce the basis set limit of interaction energies at the CCSD level of theory. [8] However, the latter misses the basis set limit of the perturbative triples contribution. The empirical (T*) scaling may be a reasonable approximation to the extrapolated value, but it does not guarantee that it reproduce the exact value.

Improvement is also observed for the non-counterpoise corrected interaction energies (Table 5.6). The discussion will be given for the imidazole·benzene complex, but the same trends hold also for the second complex. At the CCSD(T)/aug-cc-pVDZ level of theory, the difference between the corrected and non-counterpoise corrected energies exceeds 15 kJ/mol. When the explicitly-correlated CCSD(T)method is used instead of the conventional, this difference drops at around 4 kJ/mol, with or without the (T*) scaling. When interference effects are taken into account, further reduction is achieved (2.56 kJ/mol). However, this holds for the case where the aug-cc-pVDZ basis set is used for the interference-corrected MP2-F12 calculation; the corresponding difference for the cc-pVDZ-F12 basis is significantly higher (around 7 kJ/mol). The source for this is the small $\delta E_{\text{F12/cc-pVDZ-F12}}$ contribution which is not adequate to increase the interaction energies and therefore, it does not correct the overestimation from the BSSE.

On the contrary, the difference between the corrected and non-counterpoise corrected energies are the same for CCSD(T)+(F12) and CCSD(T)-INT-F12, when triple-zeta basis sets are used (around 1.4 kJ/mol). Both methods reduce the corresponding difference from the conventional CCSD(T)/aug-cc-pVTZ level (6.4 kJ/mol).

All results presented till this point (Tables 5.4, 5.5 and 5.6) have been obtained from the optimized geometries at the SCS-MP2/aug-cc-pVTZ level. However, deviations on the intermolecular distance are expected by using different levels of theory. [225] In order to examine the effect of higher

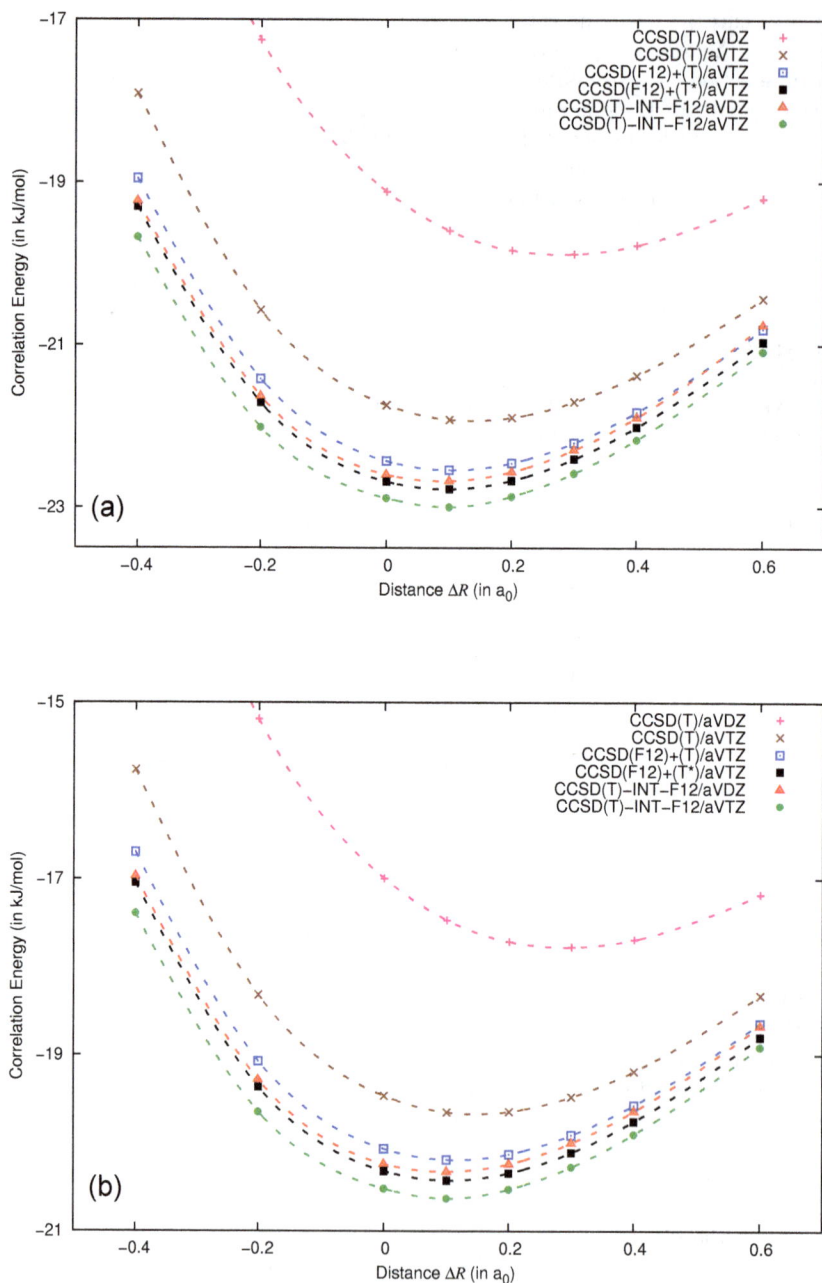

Figure 5.8: Potential energy curves of the (a) imidazole·benzene and (b) pyrrole·benzene complexes. The "CCSD(T)-INT-F12/XZ" notation corresponds to the summation of the CCSD(T)/aug-cc-pVXZ and INT-MP2-F12/cc-pVDZ-F12 energies, where $X = $ D or T. The distance ΔR is the difference of the intermolecular distance of the centers of nuclear charges from the geometry optimized at the SCS-MP2/aug-cc-pVTZ level.

levels of theory, like conventional, explicitly-correlated and interference-corrected CCSD(T), a scan over different intermolecular distances R was carried out. In these scans, the distance R between the center of mass of pyrrole (or imidazole) and the benzene plane (defined by 4 atoms) was modified at -0.4, -0.2, 0.1, 0.2, 0.3, 0.4 and 0.6 a_0 from the SCS-MP2/aug-cc-pVTZ geometry while keeping the fragments of the complexes fixed. Single-point, counterpoise corrected CCSD(T), CCSD(F12)+(T) and CCSD(T)-INT-F12 calculations were carried out on the new geometries. The aug-cc-pVXZ basis sets (X = D, T) were used for the scans. The scans are shown as potential curves in Figures 5.8(a) for the imidazole·benzene complex, Figure 5.8(b) for the pyrrole·benzene.

As it has already been discussed, the CCSD(T) interaction energies with small basis sets are inadequate to describe the strength of the noncovalent bonds. For both complexes, the curves obtained from CCSD(T) with either the double- or triple-zeta quality basis sets have a significant deviation from the curves obtained from methods which can calculate the basis set limit of coupled-cluster (explicitly-correlated or interference-corrections). The CCSD(F12) curve with the unscaled pertubative triples energies are shifted to more negative values (about 0.5 kJ/mol for both complexes). The empirical extrapolated (T*) energy to the basis set limit shifts further the potential curves at about 0.3 kJ/mol. This curve can be considered as the best estimation of the basis set limit of the CCSD(T) method, as far as all individual terms are close to their corresponding basis set limit.

The CCSD(T) curves with the energy components from the interference-corrected MP2-F12 method are very close to the best estimation from the CCSD(F12)+(T*) level of theory. In particular, energies calculated with the double-zeta basis sets almost coincide with the reference energies. The CCSD(T)-INT-F12/TZ curves are slightly lower than the best estimate for the potential curve, with a difference of about 0.2 kJ/mol.

One comment should be given about the contribution of the CABS singles to the total interaction energies. It has already been mentioned that on the SCS-MP2/aug-cc-pVTZ geometries, the contribution of this term was very small, typically less than 0.1 kJ/mol (Tables 5.4, 5.5 and 5.6). However, this correction to the HF energy becomes more important for the stretched geometries ($\Delta R \geq 0.4\,a_0$), reaching in some cases values around 0.3 kJ/mol, for the calculations with the double-zeta basis set. Therefore, the CABS singles term should be included, especially for calculations in the non-equilibrium geometries.

All the methods included in Figures 5.8(a) and (b) predict a small deviation from the equilibrium geometry optimized at the SCS-MP2/aug-cc-pVTZ level. In particular, the potential scans show a minimum at an intermolecular distance elongated by 0.1 a_0. The only exception is at the conventional CCSD(T)/aug-cc-pVDZ level of theory, which estimates the minimum distance at 0.3 a_0 longer than the SCS-MP2/aug-cc-pVTZ geometry. On the contrary, CCSD(T) with corrections from interference effects succeeds to reproduce the same equilibrium ΔR as explicitly-correlated CCSD(T) does. The use of double- or triple-zeta quality basis does not affect this result.

Based on the above, further investigation on the new optimized equilibrium geometry was carried out. The exact minima of the potential curves calculated from CCSD(F12)+(T*) have a ΔR deviations from the geometries optimized at the SCS-MP2/aug-cc-pVTZ level of 0.093 and 0.103 a_0, for the imidazole·benzene and the pyrrole·benzene complexes, respectively. In these equilibrium geometries from the higher-level method, calculations with the same methods presented in Tables 5.4, 5.5 and 5.6 were carried out. All updated results are shown in Tables 5.7 and 5.8. For both

Table 5.7: Interaction energy (ΔE_2 in kJ/mol) of the imidazole·benzene complex at its optimized geometry.

Method	aug-cc-pVDZ		aug-cc-pVTZ		aug-cc-pVQZ[a]	
	w/o CP	w/ CP	w/o CP	w/ CP	w/o CP	w/ CP
HF	-8.33	-3.69	-4.85	-3.77	-	-
MP2	-38.15	-24.12	-32.84	-26.44	-	-
SCS-MP2	-32.90	-18.71	-27.63	-20.81	-	-
DW-MP2	-36.87	-23.71	-31.30	-24.86	-	-
CCSD	-29.90	-16.78	-24.41	-18.72	-	-
CCSD(T)	-34.15	-19.57	-28.01	-21.92	-	-
HF+CABS	-4.89	-3.68	-3.96	-3.74	-3.83	-3.74
MP2-F12	-29.38	-26.72	-28.36	-27.39	-27.90	-27.45
SCS-MP2-F12	-23.75	-21.27	-22.77	-21.82	-22.32	-21.86
DW-MP2-F12	-27.80	-25.67	-26.67	-25.76	-26.20	-25.77
CCSD(F12)	-21.52	-18.85	-20.36	-19.35	-19.75	-19.39
CCSD(F12)+(T)	-25.76	-21.64	-23.97	-22.54	-23.35	-22.59
CCSD(F12)+(T*)	-26.31	-22.28	-23.96	-22.78	-23.34	-22.83
CCSD(T)-INT-F12[b]	-23.58	-22.63	-23.27	-22.84	-	-

[a] Triples corrections (T) and (T*) computed in the aug-cc-pVTZ basis.
[b] The cc-pVXZ-F12 basis sets were used.

complexes, all methods show small deviations, with the exception of the HF energies. This is not surprising, no qualitative description of noncovalent interactions can be achieved when the correlation energy is neglected and small structural changes can even affect the sign of the computed interaction energies. Counterpoise-corrected MP2 energies are smaller in absolute values for the new optimized geometry. However, the other methods, like the empirically-corrected variants of MP2, the coupled-cluster methods and the explicitly-correlated versions of them, predict a stronger amino NH··· π interaction. Only exception is the conventional dispersion-weighted MP2. The interaction energies for the SCS-MP2/aug-cc-pVTZ geometries are lower than those of Tables 5.7 and 5.8.

In line with the previous, the CCSD(T)-INT-F12 results achieves the same accuracy as the CCSD(F12)+(T*) method, where the empirically scaled pertubative triples contribution is calculated with the triple-zeta basis are added to the explicitly-correlated CCSD/aug-cc-pVQZ energy. The CCSD(T)-INT-F12 results shown in Tables 5.7 and 5.8 have been obtained by following the highly accurate computational protocol used in the S22 study (Section 5.3). This dictates the use of the cc-pVXZ-F12 basis sets, where X = D or T. For the counterpoise-corrected interaction energies, the results obtained from the double-zeta quality basis differ about 0.2 kJ/mol from the best estimates while those from the cc-pVTZ-F12 coincide with the CCSD(F12)+(T*) values. A very interesting remark is the diminish of the basis set superposition error from the CCSD(T)-INT-F12/cc-pVXZ-F12 level of theory. The energy difference of the double-zeta uncorrected values from the best estimate is at 0.75 kJ/mol for the imidazole·benzene and 0.83 for the pyrrole·benzene. This differencies are the lowest among all methods examined in this study (Tables 5.7 and 5.8). For the triple-zeta results, these deviations are further reduced. This observation is one more proof that the CCSD(T)-INT-F12 method is leading to a faster convergence of the conventional CCSD(T) method, as far as in the limit of a complete basis set the BSSE is vanishing.

Table 5.8: Interaction energy (ΔE_2 in kJ/mol) of the pyrrole·benzene complex at its optimized
geometry.

Method	aug-cc-pVDZ		aug-cc-pVTZ		aug-cc-pVQZ[a]	
	w/o CP	w/ CP	w/o CP	w/ CP	w/o CP	w/ CP
HF	-5.84	-1.06	-2.22	-1.15	-	-
MP2	-36.29	-22.05	-30.64	-24.23	-	-
SCS-MP2	-31.03	-16.63	-25.41	-18.58	-	-
DW-MP2	-34.79	-22.08	-28.76	-22.61	-	-
CCSD	-27.89	-14.60	-22.05	-16.36	-	-
CCSD(T)	-32.23	-17.48	-25.75	-19.66	-	-
HF+CABS	-2.32	-1.08	-1.34	-1.13	-1.22	-1.13
MP2-F12	-27.18	-24.44	-26.06	-25.08	-25.58	-25.13
SCS-MP2-F12	-21.51	-18.96	-20.44	-19.48	-19.97	-19.52
DW-MP2-F12	-25.21	-23.22	-23.94	-23.08	-23.44	-23.05
CCSD(F12)	-19.16	-16.44	-17.92	-16.90	-17.29	-16.94
CCSD(F12)+(T)	-23.51	-19.32	-21.63	-20.19	-20.99	-20.23
CCSD(F12)+(T*)	-24.05	-19.97	-21.62	-20.42	-20.97	-20.48
CCSD(T)-INT-F12[b]	-21.31	-20.27	-20.92	-20.48	-	-

[a] Triples corrections (T) and (T*) computed in the aug-cc-pVTZ basis.
[b] The cc-pVXZ-F12 basis sets were used.

The difference of the strength of the amino NH$\cdots\pi$ aromatic interaction of those two complexes
is of the same magnitude. For the imidazole·benzene complex, the best estimation is about -22.8
kJ/mol, while for the pyrrole·benzene it is slightly weaker, at -20.5 kJ/mol. The substitution of
the -CH group of pyrrole with a more polar nitrogen atom at the imidazole ring has an effect to
attract the electron density of the interacting hydrogen of the amino group (-NH). This makes
the specific hydrogen a better electron acceptor from the electron pool of the benzene moiety and
increases the strength of the interaction between these two molecules.

The benzene·ammonia complex (No. 18 of the S22 benchmark database) can be considered as
the simplest amino NH$\cdots\pi$ aromatic interacting system. The magnitude of the specific interaction is much smaller than the other two complexes, at -9.67 kJ/mol (S22B Revision [212]). On
the opposite, the No. 21 complex of the same S22 database, the T-shaped indole·benzene complex, has a significant stronger noncovalent bond than the one between the benzene and ammonia
moieties, at -23.5 kJ/mol. This noncovalent interaction is comparable in magnitude with the
imidazole·benzene and pyrrole·benzene interactions. However, in the case of the indole·benzene
complex, the aromatic six-membered ring of indole works as an electron attracting moiety, making in that manner the amino -NH group a better electron-acceptor and thus, increasing the
noncovalent bond.

5.5 Theoretical Studies on Hydrogen Storage in MOFs with Exposed Metal Sites

5.5.1 A short introduction on H$_2$ storage in MOFs

Alternative fuel technologies which will substitute petroleum are of high importance due to the
finite fossil fuel reserves. Among these technologies, hydrogen-based fuel cells have attracted

considerable attention. Hydrogen has two important advantages as a power generator. Firstly, its energy output per mass of hydrogen is about three times that of liquid hydrocarbons. This means that the energy content of 1 kg of oil can be produced from just 0.33 kg of hydrogen. Secondly, the sole by-product of the reaction with oxygen is water, which makes it an attractive "green" fuel. [226] Unfortunately, hydrogen is an extremely volatile gas under ambient conditions; its volumetric energy density is far too low for practical applications. For use in the automotive industry, hydrogen must be in a condensed state, either under very high pressure or cryogenically. Both have a high energy cost, greatly increase the vehicle weight and add an extra risk factor for an unsafe transport. Therefore, the target is to design new, low-cost, lightweight materials which can reversibly and rapidly store hydrogen near ambient conditions at a density equal or greater than liquid hydrogen.

Apart from the total storage capacity issue, the kinetics of release and recharging of the tank should be taken into account. Hydride formation in metal alloys, which are a storage candidate, does not offer an efficient reversibly. [156] The chemisorption association leads to bond dissociation and tight binding of the hydrogen (ranging from 50 to over 200 kJ/mol). [227] The large mass of metal hydrides is also a limitation for practical applications. On the contrary, weak dispersive interactions through physisorption circumvent the aforementioned problems. Among other parameters, physisorption correlates with surface area: greater gas uptake is favored by higher surface area. [228] Porous materials, such as metal-organic frameworks (MOFs) and certain activated carbons possess such feature.

MOFs have already been mentioned in the introduction of this Chapter (Section 5.1.2). Their framework is composed of an organic and inorganic part. The inorganic part typically contains isolated polyhedra or clusters, like in coordination chemistry, which are connected with organic linkers. [229] This inorganic part is also called secondary building unit (SBU), a term which is taken from the chemistry of zeolites. Some of the exceptional properties of MOFs are the thermal stability, permanent porosity and robustness of their frameworks.

The vast choice of different SBUs and organic linkers has as a result a broad variety of different topologies and motifs. [230, 231] Depending on the SBUs and the organic linkers which are used, different frameworks can be synthesized according to our needs. Their tunability of the active sites, pore walls and pore size offers tailor-made materials, specialized for different applications. Such applications are catalysis [232], gas adsorption, storage and separation processes [233], drug delivery and magnetic information [234].

The seminal work of Yaghi and co-workers [235] on hydrogen storage in the most famous member of this family of materials, the MOF-5, initialized the attention of MOFs as potential H_2 carriers. MOF-5 ($[OZn_4(CO_2)_6]$) is composed from octahedral Zn_4O SBUs which are bridged by six carboxylates of an organic linker (benzenedicarboxylate). Inelastic neutron scattering [236] and theoretical calculations [12, 237] have shown that the most preferable binding site of the hydrogen molecule in the pores of MOF-5 is above one face of the octahedral inorganic cluster. From studies on similar isoreticular frameworks, it was found that the binding of the H_2 molecule in this site is affected by the nature of the organic linker. [236]

5.5.2 MOFs with exposed metal sites

There are two main design principles for new functionalized porous materials. The first is the strength of the van der Waals interactions between H_2 and the frameworks; a typical isosteric heat of adsorption is in the range of 4-7 kJ/mol. For pressures up to 100 bar, an optimum heat of

adsorption should be around 20 kJ/mol. [228] The second principle on which the storage density
depends is the size of the pore. A 7 Å-wide slit pore allows exactly one layer of H_2 adsorbed
molecules on opposing surfaces and maximizes the H_2 uptake at room temperatures.

Different strategies have been proposed for increasing the H_2 adsorption enthalpies. The most
important are the incorporation of exposed metal sites in the pores, interpenetration of the material's framework for reduction of the number of large voids and hydrogen spillover. Among these
strategies, the first is the most effective and in this direction the current Section will be moved.

The term "exposed metal sites" corresponds to metals with open coordination sites. This sites can
bind H_2 molecules stronger than physisorption phenomena by forming Kubas-type complexes. [238]
The donation of the 2 σ electrons of the H_2 molecule to a vacant d-orbital forms a stable dihydrogen complex. The H_2 binds side-on to the metal and a non-classic 2-electron, 3-center bond is
formed. The metal-H_2 bond dissociation energy can reach as high as 80-90 kJ/mol.

The dihydrogen bond is elongated at around 0.89 Å, instead of 0.74 Å of the free H_2 molecule.
This is still a weaker interaction than the chemisorbed dihydrogen and can be tuned to the desired
20 kJ/mol if the favorable orbital interactions will be avoided and metal-H_2 bond is moved down
to the regime of simple charge-induced dipole interactions. [228]

Various SBU structures have been reported
in the literature which exhibit such exposed metal sites. Solvent molecules oc-
cupy these sites in the coordination sphere
of the metal centers after the synthe-
sis of the material. Solvothermal re-
moval forms the coordinatively unsaturated
centers of the metal building units. [239]
Figure 5.9 shows the most common of
them, the dinuclear (Figure 5.9a) and trinuclear (Figure 5.9b) carboxylate-bridged clusters. Black spheres represents the various metals which are forming these clusters, red and gray spheres represent the
oxygen and carbon atoms of the carboxylate bridges, respectively. while the yellow spheres are the potential H_2 binding
sites.

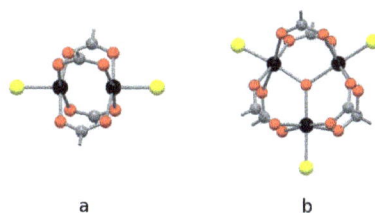

Figure 5.9: (a) Dinuclear and (b) trinuclear
carboxylate structures used as SBUs in MOFs.
Potential H_2 binding sites are shown as yellow
spheres. Black, red and gray spheres represent
metal, O and C atoms, respectively.

The bimetallic tetraboxylate paddlewheel unit $\{M_2\text{-}(O_2CR)_4\}$ is frequently formed in reactions involving Cu^{2+} and Zn^{2+} cations. In particular, MOFs which incorporate copper paddlewheels [240]
have been extensively examined for hydrogen storage. [241–247] Trinuclear oxo-centered carboxylate SBUs (Figure 5.9b) have also received much attention from synthetic chemists. They are
found in frameworks of Al^{3+} [248], Cr^{3+} [249–251], $Fe^{2+/3+}$ [252], Ni^{2+} [253] and In^{3+} [254].

The electronic structure and the magnetic properties of such SBUs (dicopper, dimanganese and
dicobalt paddlewheels, trinuclear chromium) were also the target of a recent study carried out
from the author of this Thesis. [255] The electronic structure of these complexes was investigated
prior to the calculations on interactions of the dihydrogen molecule and complexes exhibiting

uncoordinated metal centers. More specific, the ground state of these complexes was examined; do these complexes exhibit a high-spin or low-spin ground state? This was of high importance in order to study the spin contamination error introduced from DFT functionals for the low-spin cases.

The target of the current study is to examine the strength of hydrogen adsorption of such moieties. Based on the need for lighter functionalized frameworks which increase the wt % of the total hydrogen storage capacity, SBUs with light metal centers will be analyzed. Dimetallic paddlewheel structures are not included in this study, as far as they have been recently reviewed by means of wave function and density functional methods. [256] Therefore, the interaction of the dihydrogen with trinuclear clusters (Figure 5.9b) is investigated.

Oxo-centered trigonal carboxylated-bridged $\{[M_3(\mu_3\text{-O})(\text{OOCR})_6\text{L}_3]^{n+}\}$ is a family of inorganic compounds that have attracted considerable attention due to the competing exchange interactions in such systems. [255,257] The trinuclear clusters with aluminum and scandium ($M = \text{Al}^{3+}, \text{Sc}^{3+}$) were taken under consideration. For a simplification of the trigonal SBU models, the carboxylate groups are saturated with hydrogen atoms (R = H), in order to reduce the computational cost. Since every metal is in the oxidation state III, the clusters have a positive charge of +1.

The aluminum and scandium clusters are closed shell systems and show no spin polarization. The structures are optimized with the PBE functional [258] and the def2-TZVPP basis set [134, 259] and are then further optimized at the MP2/def2-TZVPP level. All metal atoms are equivalent as revealed by Mulliken and natural population analysis and by calculating the optimized $M - (\mu_3\text{-O})$ distances. For these geometries, and at the MP2/def2-TZVPP level, the interaction energy of the hydrogen molecule with the aluminum and scandium trimetalic compounds was calculated at -18.00 and -21.13 kJ/mol, respectively.

5.5.3 Modelling the strength of the H_2 adsorption

The bulky SBU trinuclear complex was modeled from a smaller, single-metal complex. In order to form a more realistic and representative model for the bigger oxo-centered trimetallic cluster, one aluminum atom with the same coordination sphere and geometry which exhibits in the original cluster was isolated (i.e. an AlO_5 fragment). The four carboxylate ligands around the metal atom were replaced by four hydroxyl groups, resulting in a negatively charged (-1) system. For this reason, the central oxygen atom was saturated with three protons (i.e. by an oxonium cation H_3O^+). Thus, a model neutral system $\{Al(OH_3)(OH)_4\}$ was constructed. The geometry of the cluster was partially optimized. All the Al–O bonds and O–Al–O angles were kept fixed and only the H–O bonds and H–O–Al angles were allowed to be optimized.

After getting the partial optimized structures of the metal cluster models, the strength of the hydrogen adsorption was studied. The complex was again partially optimized. The geometry of the $\{Al(OH_3)(OH)_4\}$ monomer was kept frozen and only the position of the H_2 molecule was free to be optimized. The final optimized geometry of the undercoordinated aluminum model with the hydrogen molecule occupying the open site of the coordination sphere of the metal is shown in Figure 5.10.1. The same procedure was followed for the H_2-scandium model cluster (Figure 5.10.2).

As far as the cluster's deformation is neglected during the geometry optimization, the interaction energy of H_2 with the smaller cluster can not be compared with the larger. For the aluminum and scandium trinuclear moieties, the interaction energies at the MP2/def2-TZVPP level are -17.99 and -21.13 kJ/mol, respectively. However, higher level calculations are possible for these

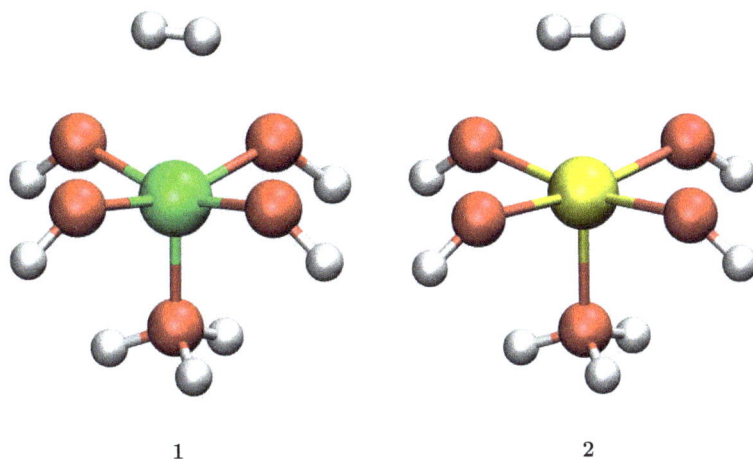

Figure 5.10: Optimized geometries of the (1) aluminum and (2) scandium undercoordinated model clusters with a hydrogen molecule occupying the open site of the coordination sphere.

models and the accuracy of various density functionals can be benchmarked. The BP86 [260], PBE [258], TPSS [261] and B3LYP [262, 263] functionals were tested. The empirical dispersion correction of Grimme [161] was also taken into consideration (DFT-D3 revision). All DFT calculations have been carried out with the TURBOMOLE program package and with the def2-TZVPP basis set.

The CCSD(T) energies at the basis set limit was used as reference interaction energies. The limit was obtained with two different approaches. The first is by extrapolating the CCSD(T) energies obtained with the double- and triple-zeta quality basis and using Helgaker's two-point formula (Eq. (2.10)). The second approach is by accounting for interference effects to the second-order perturbation theory level. The coupled-cluster and the explicitly-correlated MP2 calculations for the aluminum model complex have been done with the cc-pVXZ-F12 basis sets (X = D, T). On the contrary, for the scandium model complex, the aug-cc-pVXZ family of basis sets was used (X = D, T). The reason for this inconvenience is that no cc-pVXZ-F12 basis and corresponding cbas auxiliary basis exist for the scandium atom.

All results from density functional and wave function methods are shown in Table 5.9. These interaction energies are counterpoise corrected for the BSSE. Firstly, a short discussion on the benchmark energies from the CCSD(T) at the basis set limit will be given and based on that, conclusions on the accuracy of the different functionals will be drawn.

For both complexes, the CCSD(T) interaction energies obtained by increasing the size of the basis set, i.e. from double- to triple-zeta and from triple-zeta to the extrapolated (DT) value, show a smooth, asymptotic convergence. For the aluminum model complex (Figure 5.10.1), the interaction energies are -6.82 and -9.93 kJ/mol for the double- and triple-zeta basis, respectively, and the corresponding extrapolated value is -11.10 kJ/mol. The CCSD(T)-INT-F12 method with the double-zeta basis predicts almost the same interaction energy (-11.07 kJ/mol), while the triple-zeta result is about 0.3 kJ/mol lower (-11.37 kJ/mol). This trend between the different approaches for the CCSD(T) energies at the basis set limit agrees with the results for the S22

Table 5.9: Interaction energies (in kJ/mol) between the hydrogen molecule and the aluminum and scandium model complexes from density functional and wave function methods. All energies are counterpoise corrected. For the CCSD(T) results, XZ corresponds to cc-pVXZ-F12 basis for the aluminum complex and to aug-cc-pVXZ basis for the scandium complex (X = D, T). For the rest, the def2-TZVPP basis was used.

Level	Al-complex	Sc-complex
BP86	-1.82	-6.76
PBE	-7.43	-12.31
TPSS	-6.64	-8.67
B3LYP	-0.57	-8.34
BP86-D3	-10.81	-13.92
PBE-D3	-11.99	-16.26
TPSS-D3	-13.18	-14.09
B3LYP-D3	-8.22	-14.38
MP2-F12	-10.97	-18.84
CCSD(T)/DZ	-6.82	-16.39
CCSD(T)/TZ	-9.93	-17.89
CCSD(T)/(DT)	-11.10	-18.99
CCSD(T)-INT-F12/DZ	-11.07	-18.00
CCSD(T)-INT-F12/TZ	-11.37	-18.77
Best Estimation	-11.4	-19.0

benchmark database (Section 5.3) and for the imidazole·benzene and pyrrole·benzene Complexes (Section 5.4). Therefore, the estimated basis set limit of the noncovalent interaction between the H_2 molecule and the aluminum complex is at -11.4 kJ/mol.

On the contrary, a minor discrepancy is observed for the second model complex. From the double- and triple-zeta couple-cluster interaction energies (-16.39 and -17.89 kJ/mol, respectively) the basis set limit is estimated at around -19 kJ/mol. The CCSD(T)-INT-F12 method predicts a higher basis set limit. The aug-cc-pVDZ and aug-cc-pVTZ energies are at -18.00 and -18.77 kJ/mol, about one and 0.2 kJ/mol higher than the CCSD(T)/aug-cc-pV(DT)Z value, respectively. This deviation probably originates from the specific basis sets used for the scandium model complex. It has already been pointed out that the aug-cc-pVXZ basis sets tend to underestimate the basis set limit (Table 5.6 from Section 5.4). Therefore, the basis set limit of the noncovalent interaction between the H_2 molecule and the scandium complex is estimated at around -19.00 kJ/mol.

It should be noted that the magnitude of the scandium complex is especially important because it is close to the adsorption strength for the design of a functionalized MOF with an optimum hydrogen storage capacity (ca. 20 kJ/mol). [228] The calculation of the binding energy of the dihydrogen molecule in the available open site of the coordination sphere of the scandium atom should include also the zero-point vibrational energy. This typically increases the interaction energy (i.e. lowers the strength of the binding). On the contrary, the relaxation energy of the monomers, which has been neglected in the study of the model moiety of scandium, expects to decrease the interaction energy about 2 kJ/mol. This observation is verified from the difference between the interaction energy of the dihydrogen molecule with the trimetallic scandium compound at the MP2/def2-TZVPP level of theory. The -21.13 kJ/mol interaction energy has a difference of about 2 kJ/mol from the MP2-F12 energy for the scandium model complex. Therefore, the heat of adsorption is expected to be close to the optimum value of 20 kJ/mol.

Density functional theory without incorporating the empirical dispersion correction fails to describe the strength of the noncovalent interaction of both complexes (Table 5.9). For example, results from the PBE functional, which are closer to the estimated CCSD(T)/CBS energies among the four functionals under consideration in this Section, have a deviation of about 4 and 7 kJ/mol for the aluminum and scandium complexes, respectively. On the contrary, and as it was expected, results which include the empirical dispersion correction (mentioned as "-D3" in Table 5.9) are in a better agreement with the reference energies. Between the four DFT-D3 functionals, and for the aluminum case, the BP86-D3 and the PBE-D3 are the most accurate. The first underestimates the interaction by about 0.6 kJ/mol, while the second overestimates it by about 0.6 kJ/mol. For the scandium case, all four dispersion-corrected functionals underestimate the best estimation. From these four DFT-D3 functionals, the smallest error comes from PBE-D3 (ca. 2.7 kJ/mol). Therefore, the BP86-D3 and the PBE-D3 functionals are suggested for the study of the hydrogen adsorption of the oxo-centered trigonal carboxylated-bridged aluminum cluster and the PBE-D3 functional for the corresponding scandium cluster. It should be also mentioned that only the results obtained from the B3LYP-D3/def2-TZVPP level, even if they significantly underestimate both binding energies, almost recovers the correct difference between the best estimations of the two complexes (around 7 kJ/mol).

5.6 Summary

In this chapter, the performance of the coupled-cluster singles-and-doubles with pertubative triples method with second-order terms from the interference-corrected explicitly-correlated MP2 method on noncovalent interactions was discussed. The benefit over the conventional CCSD(T) was demonstrated firstly on the noble gas dimers and on a model acetonitrile dimer. Conclusions from these cases were used in the next Sections. Among others, the use of the Boys localization scheme and the cc-pVXZ-F12 family of basis sets was pointed out.

Interference effects lead to a fast convergence to the CCSD(T) basis set limit of the weakly interacting systems. For the acetonitrile dimer, interaction energies close to the best estimate were obtained already with a double-zeta quality basis set. This was further clarified with the S22 benchmark database. The interaction energies calculated with the CCSD(T)-INT-F12 method were compared with the S22B revision. It was found out from the statistical analysis and the deviations from the reference values that this method provides the most accurate interaction energies to date, without falling back to empirical scalings or corrections.

The strength of the amino NH$\cdots\pi$ aromatic interaction was studied in Section 5.4. In particular, two T-shaped complexes (imidazole·benzene and pyrrole·benzene) were analyzed with a variety of different post-Hartree-Fock methods, including empirically scaled approximated methods, in their conventional and explicitly-correlated variants. The basis set limit at the CCSD(T) level of these two complexes' noncovalent bond was estimated. This basis set limit was also calculated accurately from interference effects on the CCSD(T) energies, even with a double-zeta quality basis set.

In the last Section, the hydrogen binding on undercoordinated metal clusters was studied. In this application, the CCSD(T)-INT-F12 method was used as reference and different functionals of the density functional theory were tested in order to choose the most accurate functional for describing such noncovalent bonded systems. The systems under study were two small model complexes of aluminum and scandium. For the first, the BP86-D3 and PBE-D3 functionals and for the scandium complex the PBE-D3 are suggested for use in large-scale calculations of the hydrogen storage in the pores of MOFs which incorporate in their frameworks such elements with open sites in their coordination spheres.

As it has been mentioned in Chapter 4, the great advantage of the interference effects is the gain in computational time and effort over the conventional or the explicitly-correlated CCSD(T). All of them are reaching the basis set limit; conventional CCSD(T) with (very) large basis sets or with extrapolation techniques, explicitly-correlated CCSD(T) with an extra auxiliary basis and large disk space. On the contrary, the CCSD(T)-INT-F12 method reaches the CBS limit for non-covalent bonded systems by using much less disk space, smaller basis and lower computational time.

CHAPTER 6 ◼

Synopsis

A new approximate method which calculates the coupled-cluster singles-and-doubles with pertubative triples (CCSD(T)) correlation energy at the basis set limit was presented in this Thesis. The main idea behind this method is the estimation of the basis set truncation error of the CCSD(T) theory from the extrapolated second-order basis set truncation error. The second-order truncation error is estimated as the difference between the explicitly-correlated second-order Møller-Plesset perturbation theory (MP2-F12) and the conventional MP2. The second-order extrapolation depends on the asymptotic form of the correlation energy. The principle behind this asymptotic form, which achieves a faster convergence to the basis set limit, is the *interference effect*.

Interference effects are based on a pair natural orbital expansion, as shown by Petersson and co-workers. The coefficients of the pair natural orbitals are used to calculate the *interference factor*, the dominant contribution to the reduction of the truncation error. Each CCSD(T) pair energy at the basis set limit is obtained from the sum of the conventional CCSD(T) pair energy from a finite basis and the product between the interference factor and the pair energy second-order truncation error.

Almost since the foundations of quantum chemistry, **extrapolation schemes** have been a powerful tool in the toolbox of theoretical chemists. These are based either on a partial-wave, on the principal quantum number or on a natural orbital expansion. Their target is to lead to a faster convergence to the complete basis set limit. These schemes have also been the basis for a variety of different model chemistries.

An alternative way to reach a faster convergence of the energy is the utilization of **explicitly-correlated theory** (i.e. R12 or F12 methods). These theories incorporate the electronic cusp behavior explicitly into the wave function. The virtual space is augmented with explicitly correlated pair functions.

The above considerations have been taken into account in order to formulate the **interference-corrected explicitly-correlated second-order perturbation theory** (INT-MP2-F12). In this implementation, the interference factors are computed for each electron pair from the first-order wave function. The product of these factors with the second-order truncation error and summation over all electron pairs provides two independent energy terms, the F12 and the interference (INT) correction terms. When these are added to the CCSD(T) energy, a faster convergence of the correlation energy is achieved. In a shorthand notation, this method has been abbreviated as CCSD(T)-INT-F12. In addition, a dual-basis additivity scheme was also examined.

Technical considerations, like memory partitioning, orbital alignment and application of localization schemes, have extended the applicability and the accuracy of the method. The main advantages over other similar theories which estimate the CCSD(T) energy at the basis set limit are two. Firstly, the small computational cost: the interference correction is almost for free, as far as all the intermediates needed have already been precomputed, like MP2 amplitudes and localization transformation matrices. Secondly, the theory does not fall back on empirical corrections.

The performance of the CCSD(T)-INT-F12 method has been tested for **thermochemistry** and **noncovalent interactions**. For the first application, a variety of test sets were employed (106-molecule test set, AE6, BH6 and G2/97). CCSD(T)(F12) energies were used as reference computational values. For the atomization energies of the 106-molecule test set, the CCSD(T)-INT-F12/cc-pVQZ-F12 level of theory resulted in a 0.66 kJ/mol mean absolute deviation (MAD). Similar deviations were also found for the other test sets. A thermochemistry protocol was proposed based on the CCSD(T)-INT-F12 method, which has a MAD of 1.75 kJ/mol from 73 experimental atomization energies (from the active thermochemical tables database). Great accuracy was also obtained for the 6-membered barrier heights (BH6) test set and for the heats of formation of 25 molecules from the 106-molecule test set with respect to H_2, CO, CO_2, F_2 and N_2.

Noncovalent interactions were investigated by means of the CCSD(T)-INT-F12 method. As a first step, noble gas dimers and an acetonitrile dimer model geometry were studied. Conclusions from their analysis were used to accurately describe the interaction energies of the S22 benchmark database (Revision S22B). The MAD of about 0.20 kJ/mol (0.05 kcal/mol) obtained from the CCSD(T)-INT-F12/cc-pVDZ-F12 level of theory proves the level of accuracy that interference effects can introduce. The strength of the amino $NH\cdots\pi$ aromatic interaction was also studied. Different post-Hartree-Fock methods were tested and their results were compared. The CCSD(T)-INT-F12 method with double- or triple-zeta quality basis sets succeeded to accurately calculate the interaction energies of the two complexes under study, the T-shaped imidazole·benzene and pyrrole·benzene complexes. Finally, the CCSD(T)-INT-F12 method was used for benchmarking the noncovalent interaction between the hydrogen molecule and two metal complexes with exposed sites in their coordination sphere. Based on these results, density functionals were suggested for a further accurate theoretical examination of hydrogen storage in porous materials incorporating in their frameworks such undercoordinated moieties.

CHAPTER 7

Zusammenfassung (in German)

In dieser Arbeit wurde eine neue Näherungsmethode vorgestellt, die die Coupled-Cluster-Korrelationsenergie mit Einfach-, Zweifach- und störungstheoretischen Dreifachanregungen (CCSD(T)) am Basissatzlimit berechnet. Die Hauptidee hinter dieser Methode ist die Abschätzung des Fehlers in der CCSD(T)-Methode, der aus der Unvollständigkeit des Basissatzes herrührt, aus dem extrapolierten Basissatzfehler der zweiten Störungsordnung. Dieser wird aus der Differenz zwischen explizit korrelierter Møller-Plesset-Störungstheorie zweiter Ordnung (MP2-F12) und konventionellem MP2 genähert. Die Extrapolation zweiter Ordnung basiert auf der asymptotischen Form der Korrelationsenergie. Das Prinzip dahinter, das für die schnellere Konvergenz zum Basissatzlimit sorgt, ist der *Interferenzeffekt*.

Interferenzeffekte basieren auf einer natürlichen Orbitalpaar-Entwicklung, wie zuvor von Petersson und Mitarbeitern gezeigt wurde. Die Koeffizienten der natürlichen Orbitalpaar-Entwicklung werden verwendet, um den *Interferenzfaktor* zu berechnen. Dieser Faktor ist der bestimmende Beitrag zur Verringerung des Basissatzfehlers. Die CCSD(T)-Paarenergie am Basissatzlimit wurde aus der Summe der CCSD(T)-Paarenergie mit endlichem Basissatz und dem Produkt aus Interferenzfaktor und Basissatzfehler der zweiten Störungsordnung dieses Paars berechnet.

Fast seit Anbeginn der Quantenchemie sind Extrapolationsschemata ein wirkungsvolles Werkzeug in der Werkzeugkist von theoretischen Chemikern. Diese basieren entweder auf einer Partialwellen-, Hauptquantenzahl- oder natürlichen Orbitalentwicklung. Deren Ziel ist eine schnellere Konvergenz zum Basissatzlimit. Diese Schemata bilden ebenfalls die Grundlage von vielen Berechnungsvorschriften in der Quantenchemie.

Explizit korrelierte Theorien (d. h. R12- oder F12-Methoden) stellen eine Alternative dar, um eine schnellere Konvergenz zum Basissatzlimit zu erreichen. Diese Theorien beziehen das Elektronen-*Cusp*-Verhalten explizit in die Wellenfunktion mit ein. Der virtuelle Raum wird durch explizit korrelierte Paarfunktionen erweitert.

Die obigen Überlegungen wurden bei der Formulierung der **interferenzkorrigierten explizit korrelierten Störungstheorie zweiter Ordnung** (INT-MP2-F12) einbezogen. In dieser Implementierung werden die Interferenzfaktoren für jedes Elektronenpaar von der Wellenfunktion erster Ordnung berechnet. Das Produkt dieser Faktoren mit dem Basissatzfehler zweiter Ordnung sowie anschließender Summation über alle Elektronenpaare liefert zwei unabhängige Energieterme, den F12- und Interferenzkorrekturterm (INT). Wenn beide Terme zur CCSD(T)-Energie addiert werden, resultiert eine schnellere Konvergenz der Korrelationsenergie. In Kurzform wird diese Methode CCSD(T)-INT-F12 genannt. Zusätzlich wurde ein Additivitätsschema unter Verwendung zwei verschiedener Basissätze untersucht.

Durch technische Aspekte wie Speicheraufteilung, Orbitalausrichtung und Verwendung von Lokalisierungsschemata konnte die Anwendbarkeit und Genauigkeit der hier vorgestellten Methode erweitert werden. Die hauptsächlichen Vorteile gengenüber anderen ähnlichen Theorien, die das CCSD(T)-Basissatzlimit abschätzen, sind zweierlei: Erstens ist die benötigte Rechenzeit gering: die Interferenzkorrektur ist nahezu kostenlos erhältlich, weil die benötigten Intermediate bereits berechnet wurden, wie z. B. MP2-Amplituden und Transformationsmatrizen für die Lokalisierungsmethoden. Zweitens stützt diese Theorie sich nicht auf empirische Korrekturen.

Die Güte der CCSD(T)-INT-F12-Methode wurde für **Thermochemie** und **schwache Wechselwirkungen** getestet. Zunächst wurden verschiedene Testsätze verwendet (106-Molekül-Testsatz, AE6, BH6 und G2/97). Als Referenzwerte wurden CCSD(T)(F12)-Energien verwendet. Im Fall der Atomisierungsenergien des 106-Molekül-Testsatzes ergab die Methode CCSD(T)-INT-F12/cc-pVQZ-F12 eine mittlere absolute Abweichung (MAD) von 0.66 kJ/mol. Ähnliche Abweichungen ergaben sich für die anderen Testsätze. Basierend auf der Methode CCSD(T)-INT-F12/cc-pVQZ-F12 wurde eine thermochemische Berechnungsvorschrift vorgeschlagen, die einen MAD-Wert von 1.75 kJ/mol von 73 experimentellen Atomisierungsenergien (von der Datenbank ATcT, *active thermochemical tables*) erzielte. Eine hohe Genauigkeit wurde ebenfalls für die sechs Barrieren des BH6-Testsatzes und die Bildungsenthalpien von 25 Molekülen des 106-Molekül-Testsatzes relativ zu den Molekülen H_2, CO, CO_2, F_2 und N_2 gezeigt.

Schwache Wechselwirkungen wurden mit Hilfe der CCSD(T)-INT-F12-Methode untersucht. Zunächst wurden Edelgasdimere und Acetonitrildimere in Modellstrukturen betrachtet. Schlußfolgerungen aus diesen Analysen führten zu einer genauen Beschreibung der Wechselwirkungsenergien im S22-Testsatz. Der MAD-Wert von 0.20 kJ/mol (0.05 kcal/mol) der CCSD(T)-INT-F12/cc-pVDZ-F12-Methode zeigt die Stärke des Konzeptes der Interferenzkorrektur. Die Stärke der aromatischen Wechselwirkung Amino-NH$\cdots\pi$ wurde ebenfalls untersucht. Verschiedene Post-Hartree-Fock-Methoden wurden vergleichend getestet. Die CCSD(T)-INT-F12-Methode erreichte in Verbindung mit Doppel- oder Tripel-zeta-Basissätzen eine hohe Genauigkeit bei der Berechnung der Wechselwirkungsenergien zweier genauer untersuchter Komplexe, dem Imidazol· Benzol und Pyrrol·Benzol. Schließlich wurde die CCSD(T)-INT-F12-Methode verwendet, um die schwache Wechselwirkung zwischen Wasserstoffmolekülen und zwei Metallkomplexen mit ungesättigter Koordinationssphäre zu untersuchen. Basierend auf diesen Ergebnissen wurden Dichtefunktionale für Untersuchungen an ähnlichen Verbindungen vorgeschlagen, die sich zur Einlagerung von Wasserstoff in der Grundstruktur von porösen Materialien mit solchen unterkoordinierten Sphären eignen.

CHAPTER 8

Σύνοψη (in Greek)

Μια νέα προσεγγιστική μέθοδος που υπολογίζει την ενέργεια ηλεκτρονιακής συσχέτισης της μεθόδου των συζευγμένων συστάδων απλών και διπλών διεγερμένων καταστάσεων με μη επαναληπτικές τριπλές διεγέρσεις (CCSD(T)) παρουσιάζεται στην παρούσα Διατριβή. Βασική ιδέα της μεθόδου αποτελεί η εκτίμηση του σφάλματος της CCSD(T) μεθόδου, που προέρχεται από την χρήση ενός μη ολοκληρωμένου συνόλου βάσης, μέσω του αντίστοιχου $2^{ας}$-τάξεως σφάλματος, προσεγγιστικώς υπολογισμένο μέσω μοντέλων προεκβολής (extrapolation schemes). Το συγκεκριμένο σφάλμα εκτιμάται από την διαφορά μεταξύ της αναλυτικώς συσχετιζόμενης (explicitly-correlated) $2^{ας}$-τάξεως Møller-Plesset θεωρίας διαταραχών (MP2-F12) και της συμβατικής MP2. Το $2^{ας}$-τάξεως μοντέλο προεκβολής βασίζεται στην ασυμπτοτική μορφή της ηλεκτρονιακά συσχετισμένης ενέργειας. Η ταχύτερη σύγκλιση στο όριο του συνόλου βάσεως επιτυγχάνεται μέσω της *παρεμβατικής επίδρασης* (interference effect).

Οι παρεμβατικές επιδράσεις βασίζονται στο ανάπτυγμα των φυσικών τροχιακών ανά ζεύγη (pair natural orbital expansion), όπως έχει αποδείξει ο Petersson και οι συνεργάτες του. Οι συντελεστές των φυσικών τροχιακών ανά ζεύγη χρησιμοποιούνται στον υπολογισμό των *παρεμβατικών παραγόντων* (interference factors) που αποτελούν την θεμελιώδη συνεισφορά της μείωσης του σφάλματος που προέρχεται από την χρήση ενός μη ολοκληρωμένου συνόλου βάσης. Κάθε ενέργεια ζεύγους ηλεκτρονίων της CCSD(T) μεθόδου στο όριο του συνόλου βάσεως υπολογίζεται από το άθροισμα της ενέργειας ζεύγους ηλεκτρονίων της συμβατικής CCSD(T) μεθόδου από ένα πεπερασμένο σύνολο βάσης και του γινομένου μεταξύ του παρεμβατικού παράγοντα και του $2^{ας}$-τάξεως σφάλματος της αντίστοιχης ενέργειας ζεύγους ηλεκτρονίων.

Σχεδόν από την εποχή της θεμελίωσης της κβαντικής χημείας, τα **μοντέλα προεκβολής** υπήρξαν εκ των ισχυρότερων μεθόδων στην φαρέτρα των θεωρητικών χημικών. Αυτά βασίζονται στο ανάπτυγμα των μερικών κυμάτων (partial-wave), του κύριου κβαντικού αριθμού ή των φυσικών τροχιακών και στόχος τους είναι να παρέχουν ταχύτερη σύγκλιση στο όριο του συνόλου βάσεως. Αυτές οι μέθοδοι έχουν επίσης εφαρμοστεί σε διάφορα μοντέλα αυτοματοποιημένης χημείας (model chemistries).

Εναλλακτικοί τρόποι προσέγγισης της ενεργειακής σύγκλισης αποτελούν οι **αναλυτικώς συσχετιζόμενες θεωρίες** (π.χ. R12 ή F12 μέθοδοι). Αυτές οι θεωρίες συμπεριλαμβάνουν το ηλεκτρονιακό οξύ ακρότατο *αναλυτικώς* στην κυματοσυνάρτηση. Ο εικονικός χώρος στον οποίον λαμβάνουν χώρα οι διεγέρσεις των ηλεκτρονίων διευρύνεται με την προσθήκη των αναλυτικώς συσχετιζόμενων συναρτήσεων ανά ζεύγη.

Οι παραπάνω θέσεις ελήφθησαν υπ' όψη ώστε να θεμελιώσουμε την **παρεμβατικώς διορθωμένη αναλυτικώς συσχετιζόμενη $2^{ας}$-τάξεως θεωρία διαταραχών** (INT-MP2-F12). Στην συγκεκριμένη εφαρμογή της θεωρίας, οι παρεμβατικοί παράγοντες υπολογίζονται για κάθε ζεύγος ηλεκτρονίων από την $1^{ης}$-τάξεως κυματοσυνάρτηση. Αθροίζοντας τα γινόμενα των παραγόντων αυτών με το $2^{ας}$ τάξεως σφάλμα όλων των ζευγών ηλεκτρονίων, προκύπτουν δύο ανεξάρτητοι ενεργειακοί όροι, ο F12 διορθωτικός όρος και ο παρεμβατικός (INT) διορθωτικός όρος. Όταν αυτοί αθροιστούν με την ενέργεια της CCSD(T) μεθόδου, επιτυγχάνεται τάχυστη σύγκλιση της ενέργεια συσχετίσεως. Σε συντομογραφία, η μέθοδος αυτή αναφέρεται ως CCSD(T)-INT-F12.

Τεχνικές λεπτομέρειες, όπως διαχωρισμός της μνήμης RAM που καταλαμβάνουν οι μήτρες που χρησιμοποιούνται κατά τον υπολογισμό των παρεμβατικών παραγόντων, παραλληλισμός (alignment) των άλφα και βήτα τροχιακών, καθώς και η χρησιμοποίηση εντοπισμένων τροχιακών (localized orbitals) διεύρυναν την εφαρμοσιμότητα και την ακρίβεια της μεθόδου. Δύο είναι τα βασικά πλεονεκτήματα σε σχέση με άλλες παρόμοιες μεθόδους που εκτιμούν το όριο του συνόλου βάσεως. Κατά πρώτον, το μικρό υπολογιστικό κόστος: η παρεμβατική διόρθωση δε χρειάζεται επιπρόσθετο υπολογιστικό χρόνο, διότι όλοι οι ενδιάμεσοι όροι έχουν προϋπολογιστεί, όπως, επί παραδείγματι οι όροι της $1^{ης}$-τάξεως κυματοσυνάρτησης και οι μήτρες μετασχηματισμού των εντοπισμένων τροχιακών. Κατά δεύτερον, η θεωρία δεν στηρίζεται σε εμπειρικούς παράγοντες ή εμπειρικές διορθώσεις.

Η CCSD(T)-INT-F12 θεωρία χρησιμοποιήθηκε σε δύο διαφορετικές εφαρμογές, στη **θερμοχημεία** και στη μελέτη **ασθενών αλληλεπιδράσεων**. Για την πρώτη εφαρμογή εξετάστηκαν διαφορετικά δοκιμαστικά σύνολα μορίων (106-molecule test set, AE6, BH6 and G2/97). Οι CCSD(T)(F12) ενέργειες χρησιμοποιήθηκαν ως υπολογιστικές τιμές αναφοράς. Για τις ενέργειες ατομοποίησης του δοκιμαστικού συνόλου των 106 μορίων, το CCSD(T)-INT-F12/cc-pVQZ-F12 επίπεδο θεωρίας έδωσε 0.66 kJ/mol μέση απόλυτη απόκλιση (mean absolute deviation - MAD). Παρεμφερείς αποκλί- σεις υπολογίστηκαν και για τα υπόλοιπα σύνολα μορίων. Από το πρωτόκολο θερμοχημείας που προτάθηκε βασισμένο στην CCSD(T)-INT-F12 μέθοδο επετεύχθη μέση απόλυτη απόκλιση της τάξεως των 1.75 kJ/mol σε σχέση με 73 πειραματικές ενέργειες ατομοποίησης. Πολύ καλή ακρίβεια επίσης επετεύχθη για το δοκιμαστικό σύνολο των 6 ενεργειακών φραγμάτων (barrier heights) και για τις ενέργειες σχηματισμού 25 μορίων σε σχέση με τα H_2, CO, CO_2, F_2 και N_2.

Η δεύτερη εφαρμογή σχετίζεται με την μελέτη ασθενών αλληλεπιδράσεων. Ως πρώτο βήμα μελετήθηκαν τα διμερή των ευγενών αερίων και ένα μοντέλο του διμερούς του ακετονιτριλίου. Τα συμπεράσματα που εξήχθησαν χρησιμοποιήθηκαν έτσι ώστε να επιτευχθεί ακριβής περιγραφή των ενεργειών αλληλεπίδρασης της S22 βάσης δεδομένων. Η μέση απόλυτη απόκλιση των 0.20 kJ/mol (0.05 kcal/mol) από τις ενέργειες αναφοράς, όπως υπολογίστηκαν από το CCSD(T)-INT-F12/cc-pVDZ-F12 επίπεδο θεωρίας, αποτελεί απόδειξη της ακρίβειας που μπορεί να επιτευχθεί μέσω των παρεμβατικών επίδρασεων. Η ισχύς της άμινο NH$\cdots\pi$ αρωματικής αλληλεπίδρασης μελετήθηκε επίσης. Διαφορετικές post-Hartree-Fock μέθοδοι δοκιμάστηκαν και τα αποτελέσματά τους συνεκρίθησαν μεταξύ τους. Τα αποτελέσματα που ελήφθησαν από την CCSD(T)-INT-F12 μέθοδο με διπλή- ή τριπλή-ζήτα σύνολο βάσης περιγράφουν με μεγάλη ακρίβεια τις ενέργειες αλληλεπίδρασης των δύο συμπλεγμάτων που μελετήθηκαν, (ιμιδαζόλιο·βενζόλιο και πυρρόλιο·βενζόλιο σε Τ-διαμόρφωση). Τέλος, η CCSD(T)-INT-F12 μέθοδος χρησιμοποιήθηκε ως αναφορά για συγκριτική αξιολόγηση διαφορετικών συναρτησιακών που μπορούν να περιγράψουν με ακρίβεια ασθενείς αλληλεπιδράσεις μεταξύ του μορίου του υδρογόνου και δύο μεταλλικών συμπλόκων με ελεύθερες θέσεις στην σφαίρα σύνταξής τους. Με τα προτεινόμενα συναρτησιακά, υπολογισμοί σε μεγάλη κλίμακα μπορούν να επιτευχθούν για την θεωρητική μελέτη αποθήκευσης υδρογόνου σε πορώδη νανοδομημένα υλικά που περιέχουν τα συγκεκριμένα σύμπλοκα.

Appendices

Acronyms

AE6	six-membered Atomization Energy test set
AO	Atomic Orbital
APNO	Atomic Pair Natural Orbital
ATcT	Active Thermochemical Tables
BH6	six-membered reaction Barrier Height test set
BSSE	Basis Set Superposition Error
CABS	Complementary Auxiliary Basis Set
CBS	Complete-Basis-Set
CCSD	Coupled-Cluster Singles-and-Doubles
CCSD(T)	Coupled-Cluster Singles-and-Doubles with pertubative Triples
CCSDT	Coupled-Cluster Singles-Doubles-and-Triples
ccCA	correlation-consistent Composite Approach
cc-pVXZ	correlation-consistent polarized Valence X-Zeta basis set
cc-pCVXZ	correlation-consistent polarized Core-Valence X-Zeta basis set
CEPA	Coupled-Electron Pair Approximation
CI	Configuration-Interaction
CP	Counterpoise correction
CV	Core-Valence
DFT	Density Functional Theory
DW	Dispersion-Weighted
INT	second-order Interference energy correction
INT-MP2-F12	Interference-corrected explicitly-correlated MP2
fc	frozen-core
FCI	Full Configuration-Interaction
GVB	Generalized Valence Bond
Gn	Gaussian-n model chemistries
HEAT	High-accuracy Extrapolated *ab initio* Thermochemistry
HF	Hartree-Fock

LB	Large Basis set (for additivity scheme)
LMO	Localized Molecular Orbital
MAD	Mean-Absolute Deviation
MCSCF	Multi-Configuration Self-Consistent Field
MO	Molecular Orbital
MOF	Metal-Organic Framework
MP2	Second-Order Møller-Plesset perturbation theory
MP2-F12	Explicitly-correlated MP2
NO	Natural Orbital
PNO	Pair Natural Orbital
QCI	Quadratic Configuration-Interaction
QCISD(T)	QCI Singles-and-Doubles and pertubative Triples
RI	Resolution-of-the-Identity
RMS	Root-Mean-Square
ROCBS	Restricted Open-shell Complete-Basis-Set model chemistry
ROHF	Restricted Open-shell Hartree-Fock
SAPT	Symmetry-Adapted Perturbation Theory
SB	Small Basis set (for additivity scheme)
SBU	Secondary Building Unit
SCF	Self-Consistent-Field
SCS-MP2	Spin-Component-Scaled MP2
SOS-MP2	Scalled-Opposite-Spin MP2
SVD	Singular-Value-Decomposition method
S22	Structures 22 benchmark database
UHF	Unrestricted Hartree-Fock
Wn	Weizmann-n model chemistries
ZIF	Zeolitic Imidazole Framework

On the Asymptotic Convergence of the PNO Expansions

B.1 The Helium case

In Ref. [13] and [51], the derivation of the interference factors from the pair natural orbital (PNO) expansion is given. For sake of completeness, it is also included in the current Thesis.

For N different states, the approximate wave functions are:

$$\psi_k^{(0)} = \sum_{\mu=1}^{N} C_\mu^k \Phi_\mu, \qquad k = 1, 2, 3, \ldots, N. \tag{B.1}$$

The orbitals can be selected either as PNOs or as multi-configuration self-consistent field (MCSCF) orbitals to variationally optimize any root of this configuration-interaction (CI) problem. This PNO or MCSCF root is defined as the zero state of the zero-order Hamiltonian:

$$\psi_0^{(0)} = \sum_{\mu=1}^{N} C_\mu \Phi_\mu, \qquad |C_{n+1}| \leq |C_n|, \tag{B.2}$$

and assume that the configurations are ordered with decreasing occupation numbers. The remaining states of $\hat{\mathcal{H}}^{(0)}$ are defined as single PNO configurations:

$$\psi_k^{(0)} = \Phi_k, \qquad k = N+1, N+2, \ldots, \infty. \tag{B.3}$$

The error in the energy calculated from the multi-configuration wave function is then

$$E - E(N) = E^{(2)} + E^{(3)} + \ldots, \tag{B.4}$$

where $E^{(2)}$ corresponds to the energy obtained from second-order perturbation theory, $E^{(3)}$ from third-order perturbation theory [39] and so forth:

$$E^{(2)} = \sum_{k=N+1}^{\infty} \frac{|\langle \psi_0^{(0)}|\hat{\mathcal{H}}|\psi_k^{(0)}\rangle|^2}{E_0^{(0)} - H_k^{(0)}}, \tag{B.5}$$

$$E^{(3)} = \sum_{k=N+1}^{\infty} \sum_{l>k}^{\infty} \frac{2\langle \psi_0^{(0)}|\hat{\mathcal{H}}|\psi_k^{(0)}\rangle\langle \psi_k^{(0)}|\hat{\mathcal{H}}|\psi_l^{(0)}\rangle\langle \psi_l^{(0)}|\hat{\mathcal{H}}|\psi_0^{(0)}\rangle}{(H_0^{(0)} - H_k^{(0)})(H_0^{(0)} - H_l^{(0)})}, \tag{B.6}$$

$$\vdots \tag{B.7}$$

To simplify notation, matrix elements between different configurations will be written as

$$H_{\mu\nu} = \langle \Phi_\mu|\hat{\mathcal{H}}|\Phi_\nu\rangle. \tag{B.8}$$

and the energy differences of the denominators as

$$\Delta E_\lambda = E(N) - E_\lambda \tag{B.9}$$

By substituting Eqs. (B.1), (B.2), (B.3), (B.8) and (B.9) to Eq. (B.4), while retaining terms through third order, a perturbation expansion is being obtained for the PNO correlation energy:

$$E - E(N) = \sum_{\mu=1}^{N} \sum_{\nu=1}^{N} C_\mu C_\nu \left[\sum_{\lambda=N+1}^{\infty} \frac{H_{\mu\lambda}}{\Delta E_\lambda} \left(H_{\lambda\nu} + \sum_{\sigma>\lambda}^{\infty} \frac{2H_{\lambda\sigma} H_{\sigma\nu}}{\Delta E_\sigma} + \ldots \right) \right]. \tag{B.10}$$

The dominant factors to the PNO truncation error can be isolated by partitioning of Eq. (B.10) into three terms. The partition can be done in two different ways, corresponding to the two extremes of behavior for the CI coefficients.

As first case, a closed shell system is considered, where the coefficient of the first NO configuration is much larger than the coefficients of the remaining NOs. It is then reasonable to partition the terms of Eq. (B.10) into those with $C_\mu C_\nu$ equal to C_1^2, $C_1 C_{\mu \neq 1}$ and $C_{\mu \neq 1} C_{\nu \neq 1}$. The terms which contains the C_1^2 factor would tend to be the largest. However, this term by itself cannot explain the fact that the energy will be lower by adding a new configuration to the CI expansion on which already other configurations have been added.

The second extreme in the behavior of the CI coefficients is the degenerate open-shell case, in which several configurations have similar coefficients (e.g. the homolytic cleavage of a bond). The NOs of these nearly degenerate configurations will necessarily be similar in size and therefore have comparable matrix elements $H_{\mu\lambda}$ with the higher NOs. Therefore, a reasonable partitioning can be done:

$$E - E(N) = \left(\sum_{\mu=1}^{N} C_\mu \right)^2 \left[\sum_{\lambda=N+1}^{\infty} \frac{H_{1\lambda}}{\Delta E_\lambda} \left(H_{\lambda 1} + \sum_{\sigma>\lambda}^{\infty} \frac{2H_{\lambda\sigma} H_{\sigma 1}}{\Delta E_\sigma} + \ldots \right) \right] \tag{B.11}$$

$$+ 2\left(\sum_{\mu=1}^{N} C_\mu \right) \sum_{\nu=2}^{N} C_\nu \left\{ \sum_{\lambda=N+1}^{\infty} \frac{H_{1\lambda}}{\Delta E_\lambda} \left[(H_{\lambda\nu} - H_{1\lambda}) + \sum_{\sigma\neq\lambda}^{\infty} \frac{H_{\lambda\sigma}(H_{\sigma\nu} - H_{\sigma 1})}{\Delta E_\sigma} + \ldots \right] \right\}$$

$$+ \sum_{\mu=2}^{N} \sum_{\nu=2}^{N} C_\mu C_\nu \left\{ \sum_{\lambda=N+1}^{\infty} \frac{(H_{\mu\lambda} - H_{1\lambda})}{\Delta E_\lambda} \left[(H_{\lambda\nu} - H_{1\lambda}) + \sum_{\sigma>\lambda}^{\infty} \frac{2H_{\lambda\sigma}(H_{\sigma\nu} - H_{\sigma 1})}{\Delta E_\sigma} + \ldots \right] \right\}.$$

The first term is consistent with the energy lowering from the other configurations and reduces to the C_1^2 terms of the first case if $C_1 \sim 1$. Nyden and Petersson [13] showed for the ground state of the H_2 molecule that the error introduced by neglecting the rest terms of Eq. (B.11) is smaller than the respecting from second-order perturbation theory or a MCSCF calculation including 5 configurations. However, the main conclusion remains that this is by far the dominant term and it recovers the correlation energy at the dissociation limit. Further simplification of Eq. (B.11) can be achieved by neglecting higher orders of the perturbation series:

$$\lim_{\substack{N\to\infty \\ C_{\mu\neq 1}\to 0}} \delta E(N) \cong \left(\sum_{\mu=1}^{N} C_\mu \right)^2 \left(\sum_{\lambda=N+1}^{\infty} \frac{H_{1\lambda} H_{\lambda 1}}{\Delta E_\lambda} \right). \tag{B.12}$$

This asymptotic form gives the relationship between the total multi-configuration error and the component of the second-order single configuration perturbation energy from $\lambda > N$. The last term of Eq. (B.12) corresponds to the second-order perturbation expansion of Schwarz (Section

2.1.1, Eq. (2.4)). Nyden and Petersson [13] have also proved that substitution of the $(\ell + \frac{1}{2})^{-4}$ part of this expansion with a term depending on the number of NOs leads to an expression which is valid for the convergence of PNOs expansions

$$\lim_{N \to \infty} \delta E(N) \cong \left(\sum_{\mu=1}^{N} C_\mu \right)^2 (-225/4608)N^{-1} . \tag{B.13}$$

The only restriction is on the number of NOs which should be included. N must composed from complete shells, according to the principal quantum number n: $N = 1$ for 1s, $N = 5$ for 1s2s2p, $N = 14$ for 1s2s2p3s3p3d shells of NOs, and so on.

In order to empirically reproduce the observed behavior for a small number of N, a parameter δ was introduced, without affecting the asymptotic behavior.

$$\lim_{N \to \infty} \delta E(N) \cong \left(\sum_{\mu=1}^{N} C_\mu \right)^2 (-225/4608)(N + \delta)^{-1} . \tag{B.14}$$

δ is a system-depending parameter. For the helium atom and with a fixed value of $\delta = 0.363$, an accuracy of 0.00005 Hartree was achieved.

B.2 Extension to multi-electron systems

Extension of the above for a generalization to a multi-electron system needs another important consideration. The intraorbital pair energies e_{ii} and the $\alpha\beta$- and $\sigma\sigma$-interorbital pair energies ($^{\alpha\beta}e_{ij}$ and $^{\sigma\sigma}e_{ij}$) should be considered separately, since each type of pair interaction shows a different asymptotic convergence as the basis set increases. These three cases have been examined separately from Petersson $et\ al.$ [51] for the test case of the He$_n$ system, but their results are transferable to any multi-electron system.

a. $\alpha\beta$-intraorbital pair energies

The asymptotic limit of the second-order pair correlation energy (Eq. (2.14)) is

$$\lim_{N_{ii} \to \infty} \delta e_{ii}^{(2)} = (-225/4608)(N_{ii} + \delta_{ii})^{-1} , \tag{B.15}$$

and the infinite-order pair energy (CI-limit) (Eq. (2.15)) is

$$\lim_{N_{ii} \to \infty} \delta e_{ii} = \left(\sum_{\mu=1}^{N_{ii}} C_{\mu_{ii}} \right)^2 (-225/4608)(N_{ii} + \delta_{ii})^{-1} . \tag{B.16}$$

The asymptotic convergence of the higher-order (n-order) contributions to the intraorbital pair energies are given by the difference between Eqs. (B.15) and (B.16).

$$\lim_{N_{ii} \to \infty} \delta \left(\sum_{n=3}^{\infty} e_{ii}^{(n)} \right) = \left[1 - \left(\sum_{\mu=1}^{N_{ii}} C_{\mu_{ii}} \right)^2 \right] (-225/4608)(N_{ii} + \delta_{ii})^{-1} . \tag{B.17}$$

b. $\alpha\beta$-interorbital pair energies

The asymptotic convergence of the second-order $\alpha\beta$-interorbital pair energies has the form

$$\lim_{N_{ij}\to\infty} \delta^{\alpha\beta} e_{ij}^{(2)} = {}^{\alpha\beta} f_{ij}(-225/4608)(N_{ij} + {}^{\alpha\beta}\delta_{ij})^{-1}, \tag{B.18}$$

where $0 \leq {}^{\alpha\beta} f_{ij} \leq 1$ is the overlap prefactor and it is calculated from

$$^{\alpha\beta} f_{ij} = |S|_{ij}^2, \tag{B.19}$$

where the $|S|_{ij}$ is the absolute overlap integral

$$|S|_{ij} = \int |\phi_i(\vec{r})\phi_j(\vec{r})| d\vec{r}. \tag{B.20}$$

It should be clear that $|S|_{ij}$ is the integral over the *absolute value* of the product of the occupied orbitals and not the absolute value of the overlap integral. It requires a numerical integration. In analogy with Eq. (B.17), the higher-order of the perturbation expansion can be estimated from the basis set truncation

$$\lim_{N_{ij}\to\infty} \delta\left(\sum_{n=3}^{\infty} {}^{\alpha\beta} e_{ij}^{(n)}\right) = {}^{\alpha\beta} f_{ij}\left[1 - \left(\sum_{\mu=1}^{N_{ij}} C_{\mu ij}\right)^2\right](-225/4608)(N_{ij} + {}^{\alpha\beta}\delta_{ij})^{-1}. \tag{B.21}$$

One additional generalization can be done by treating both $\alpha\beta$-intra- and interorbital pairs in the same manner, where $|S|_{ii} = 1$.

c. $\sigma\sigma$-interorbital pair energies

Schwarz [15] had noted that the Fermi-hole present in the wave function for *same-spin* $\sigma\sigma$-pairs should lead to an l^{-6} asymptotic form for the second-order contribution from angular momentum l. This gives a total basis set error of the form al^{-5}. Conversion to the total number of N_{ij} PNO configurations gives a total basis set error of the form $b(N_{ij} + \delta_{ij})^{-5/3}$. Therefore, in analogy with the $\alpha\beta$-pairs, the basis set limit of the second-order pair correlation energy is estimated from

$$\lim_{N_{ij}\to\infty} \delta^{\sigma\sigma} e_{ij}^{(2)} = {}^{\sigma\sigma} f_{ij}(-225/4608)(N_{ij} + {}^{\sigma\sigma}\delta_{ij})^{-5/3}, \tag{B.22}$$

where the overlap factor ${}^{\sigma\sigma} f_{ij} \neq {}^{\alpha\beta} f_{ij}$ is set equal to [264]

$$^{\sigma\sigma} f_{ij} = 2|S|_{ij}^2\left(\frac{1 - |S|_{ij}^2}{1 + |S|_{ij}^2}\right), \tag{B.23}$$

The purpose of the prefactors ${}^{\alpha\beta} f_{ij}$ and ${}^{\sigma\sigma} f_{ij}$ is to damp the extrapolated truncation errors for spatially distant pairs of orbitals. The $(-5/3)$ exponent implies also that the $\sigma\sigma$-interorbital pair energies converges faster than the $\alpha\beta$-pairs. By analogy with Eq. (B.21), the higher-order $\sigma\sigma$-pair convergence has the form

$$\lim_{N_{ij}\to\infty} \delta\left(\sum_{n=3}^{\infty} {}^{\sigma\sigma} e_{ij}^{(n)}\right) = {}^{\sigma\sigma} f_{ij}\left[1 - \left(\sum_{\mu=1}^{N_{ij}} C_{\mu ij}\right)^2\right](-225/4608)(N_{ij} + {}^{\sigma\sigma}\delta_{ij})^{-5/3}. \tag{B.24}$$

Total correlation energies of selected atoms and molecules

The total energies of selected atoms and molecules are presented (from Ref. [41]) calculated with the cc-pVTZ-F12 basis. Note that there is a mistake on the original Table 3 of Ref. [41]. The $E_{Int/A}$ and $E_{Int/B}$ of the fluorine atom have been erroneously calculated. The source of this error was a false multiplication between the interference factor and the e_{ij}^{F12} for two electron pairs. On the present Table C.1, the corrected values have been included.

System	δE_{MP2}	$\delta E_{F12/A}$	$\delta E_{Int/A}$	$\delta E_{F12/B}$	$\delta E_{Int/B}$	$\delta E_{CCSD(T)}$	$\delta E_{CCSD(T)(F12)}$
C	−70.71	−6.05	2.75	−5.52	2.51	−92.42	−95.18
N	−101.66	−9.19	3.43	−8.45	3.16	−119.53	−124.75
O	−152.29	−18.82	5.91	−17.32	5.44	−171.79	−184.40
F	−214.87	−26.93	7.42	−24.94	6.87	−230.46	−249.93
CH$_4$	−201.77	−18.58	8.26	−16.91	7.52	−228.31	−239.09
NH$_3$	−243.01	−22.95	9.01	−20.97	8.24	−262.86	−277.24
H$_2$O	−273.16	−28.81	9.98	−26.45	9.17	−286.36	−305.87
C$_2$H$_2$	−320.30	−26.77	11.28	−24.25	10.22	−347.85	−363.19
C$_2$H$_4$	−344.65	−30.08	12.69	−27.35	11.54	−383.52	−400.97
C$_2$H$_6$	−377.91	−34.17	14.39	−31.08	13.10	−422.36	−442.36
HCN	−359.18	−29.62	11.61	−26.98	10.58	−380.09	−397.80
HF	−288.60	−32.37	9.51	−29.91	8.79	−297.11	−320.46
N$_2$	−390.94	−32.28	11.99	−29.53	10.98	−406.63	−426.68
CO	−371.68	−34.36	12.15	−31.52	11.15	−390.95	−413.12
CH$_3$OH	−443.53	−44.19	15.53	−40.41	14.22	−476.65	−505.15
CH$_3$F	−454.79	−47.71	15.39	−43.81	14.14	−484.26	−516.56
H$_2$O$_2$	−519.36	−53.36	17.89	−48.95	16.42	−545.27	−581.09
N$_2$O	−670.82	−58.03	19.33	−53.13	17.72	−684.58	−722.16
F$_2$	−555.17	−58.94	16.81	−54.44	15.54	−577.62	−619.49
O$_3$	−812.85	−71.60	22.82	−65.74	20.98	−826.57	−873.62

Table C.1: Valence-shell correlation energies and interference corrections of selected atoms and molecules, obtained in the cc-pVTZ-F12 basis. All energies are in mE_h. δE_{MP2} and $\delta E_{CCSD(T)}$ are the conventional valence-shell correlation energies. $\delta E_{F12/A}$ and $\delta E_{F12/B}$ are the estimates for the basis-set truncation error as obtained at the MP2-F12/A and MP2-F12/B levels, respectively, and $\delta E_{INT/A}$ and $\delta E_{INT/B}$ are the corresponding interference corrections. $\delta E_{CCSD(T)(F12)}$ is the CCSD(T)(F12) valence-shell correlation energy.

Statistics of the Additivity Scheme

The correlation energy statistics of the atomization energies of the 106-molecule test set as being calculated from the additivity scheme are shown in the next tables. As reference correlation energy, the CCSD(T)(F12)/def2-QZVPP level of theory was used. These tables include all the results from the different basis set combinations taken into account in this thesis, the explicitly-correlated MP2 approximations, the level of the interference-corrected MP2-F12 ($F^{ij} = 1$ or explicit calculation of the interference factor) and the use or not of the empirical scaling of the perturbative triples, named as (T*). All results are in kJ/mol. For further discussion, see Section 4.1.5.

Method	mean error	MAD	RMS	Max Error	Molecule
		cc-pVDZ-F12			
UCCSD(T)+F12/A	27.65	27.65	28.92	57.6	104.N_2O_4
UCCSD(T)+F12/B	12.37	12.37	13.09	27.9	104.N_2O_4
UCCSD(T*)+F12/A	42.80	42.80	45.42	109.5	104.N_2O_4
UCCSD(T*)+F12/B	26.31	26.31	28.42	75.5	104.N_2O_4
UCCSD(T)-INT-F12/A	-15.10	15.22	16.58	-32.4	70.C_4H_4
UCCSD(T)-INT-F12/B	-26.55	26.55	28.21	-52.6	70.C_4H_4
UCCSD(T*)-INT-F12/A	0.05	8.22	10.50	36.9	103.N_2O_3
UCCSD(T*)-INT-F12/B	-12.62	13.94	16.40	-37.4	70.C_4H_4
		cc-pVTZ-F12			
UCCSD(T)+F12/A	20.50	20.50	21.32	40.4	104.N_2O_4
UCCSD(T)+F12/B	15.24	15.24	15.96	30.6	104.N_2O_4
UCCSD(T*)+F12/A	35.10	35.10	37.26	90.2	104.N_2O_4
UCCSD(T*)+F12/B	29.47	29.47	31.53	79.1	104.N_2O_4
UCCSD(T)-INT-F12/A	-20.36	20.36	22.00	-43.0	70.C_4H_4
UCCSD(T)-INT-F12/B	-24.27	24.27	25.97	-49.2	70.C_4H_4
UCCSD(T*)-INT-F12/A	-5.76	9.76	11.83	-27.0	70.C_4H_4
UCCSD(T*)-INT-F12/B	-10.04	12.09	14.50	-33.6	70.C_4H_4
		cc-pVQZ-F12			
UCCSD(T)+F12/A	17.28	17.28	18.02	34.1	104.N_2O_4
UCCSD(T)+F12/B	15.57	15.57	16.28	30.9	104.N_2O_4
UCCSD(T*)+F12/A	31.73	31.73	33.81	83.4	104.N_2O_4
UCCSD(T*)+F12/B	29.89	29.89	31.94	79.7	104.N_2O_4
UCCSD(T)-INT-F12/A	-22.70	22.70	24.44	-47.2	70.C_4H_4
UCCSD(T)-INT-F12/B	-23.97	23.97	25.74	-49.1	70.C_4H_4
UCCSD(T*)-INT-F12/A	-8.26	11.07	13.37	-31.5	70.C_4H_4
UCCSD(T*)-INT-F12/B	-9.65	11.95	14.32	-33.5	70.C_4H_4

Table D.1: Small Basis: **cc-pVDZ**

Method	mean error	MAD	RMS	Max Error	Molecule
cc-pVDZ-F12					
UCCSD(T)+F12/A	21.08	21.08	22.34	41.9	$104.N_2O_4$
UCCSD(T)+F12/B	5.80	5.84	6.28	12.1	$104.N_2O_4$
UCCSD(T*)+F12/A	28.15	28.15	29.79	65.1	$104.N_2O_4$
UCCSD(T*)+F12/B	11.56	11.56	12.33	31.2	$104.N_2O_4$
UCCSD(T)-INT-F12/A	-1.98	2.59	3.05	-9.1	$67.C_3O_2$
UCCSD(T)-INT-F12/B	-12.78	12.78	13.31	25.7	$104.N_2O_4$
UCCSD(T*)-INT-F12/A	5.08	5.32	6.82	23.6	$103.N_2O_3$
UCCSD(T*)-INT-F12/B	-7.03	7.08	7.69	-15.8	$70.C_4H_4$
cc-pVTZ-F12					
UCCSD(T)+F12/A	13.93	13.93	14.55	24.7	$104.N_2O_4$
UCCSD(T)+F12/B	8.67	8.67	9.09	14.9	$104.N_2O_4$
UCCSD(T*)+F12/A	20.40	20.40	21.37	45.8	$104.N_2O_4$
UCCSD(T*)+F12/B	14.73	14.73	15.47	34.7	$104.N_2O_4$
UCCSD(T)-INT-F12/A	-6.94	6.94	7.35	-16.2	$104.N_2O_4$
UCCSD(T)-INT-F12/B	-10.57	10.57	11.08	-23.2	$104.N_2O_4$
UCCSD(T*)-INT-F12/A	-0.48	2.46	3.12	12.4	$103.N_2O_3$
UCCSD(T*)-INT-F12/B	-4.51	4.80	5.42	-12.2	$70.C_4H_4$
cc-pVQZ-F12					
UCCSD(T)+F12/A	10.71	10.71	11.21	18.4	$104.N_2O_4$
UCCSD(T)+F12/B	9.00	9.00	9.43	15.1	$104.N_2O_4$
UCCSD(T*)+F12/A	17.00	17.00	17.82	39.0	$104.N_2O_4$
UCCSD(T*)+F12/B	15.15	15.15	15.89	35.3	$104.N_2O_4$
UCCSD(T)-INT-F12/A	-9.16	9.16	9.63	-20.8	$104.N_2O_4$
UCCSD(T)-INT-F12/B	-10.34	10.34	10.85	-23.1	$104.N_2O_4$
UCCSD(T*)-INT-F12/A	-2.88	3.48	4.13	-10.1	$70.C_4H_4$
UCCSD(T*)-INT-F12/B	-4.19	4.52	5.16	-12.0	$70.C_4H_4$

Table D.2: Small Basis: **cc-pVTZ**

Method	mean error	MAD	RMS	Max Error	Molecule
	cc-pVDZ-F12				
UCCSD(T)+F12/A	18.74	18.74	20.00	37.2	$104.N_2O_4$
UCCSD(T)+F12/B	3.46	3.54	3.98	7.5	$104.N_2O_4$
UCCSD(T*)+F12/A	22.39	22.39	23.82	49.3	$104.N_2O_4$
UCCSD(T*)+F12/B	5.81	5.81	6.35	15.5	$104.N_2O_4$
UCCSD(T)-INT-F12/A	5.24	5.31	6.11	12.3	$103.N_2O_3$
UCCSD(T)-INT-F12/B	-5.15	5.15	5.38	-11.5	$104.N_2O_4$
UCCSD(T*)-INT-F12/A	8.88	8.89	9.90	22.1	$103.N_2O_3$
UCCSD(T*)-INT-F12/B	-2.80	2.81	3.00	-5.7	$70.C_4H_4$
	cc-pVTZ-F12				
UCCSD(T)+F12/A	11.59	11.59	12.16	20.0	$104.N_2O_4$
UCCSD(T)+F12/B	6.32	6.32	6.69	10.8	$31.CH_4N_2O$
UCCSD(T*)+F12/A	14.65	14.65	15.35	30.0	$104.N_2O_4$
UCCSD(T*)+F12/B	8.98	8.98	9.43	19.0	$104.N_2O_4$
UCCSD(T)-INT-F12/A	0.47	0.92	1.23	4.2	$39.C_2HF_3$
UCCSD(T)-INT-F12/B	-2.98	2.98	3.19	-9.0	$104.N_2O_4$
UCCSD(T*)-INT-F12/A	3.53	3.53	4.02	11.2	$103.N_2O_3$
UCCSD(T*)-INT-F12/B	-0.33	0.84	1.06	4.3	$103.N_2O_3$
	cc-pVQZ-F12				
UCCSD(T)+F12/A	8.37	8.37	8.81	14.5	$31.CH_4N_2O$
UCCSD(T)+F12/B	6.65	6.65	7.03	11.5	$31.CH_4N_2O$
UCCSD(T*)+F12/A	11.24	11.24	11.78	23.2	$104.N_2O_4$
UCCSD(T*)+F12/B	9.39	9.39	9.85	19.5	$104.N_2O_4$
UCCSD(T)-INT-F12/A	-1.61	1.63	1.91	-6.5	$104.N_2O_4$
UCCSD(T)-INT-F12/B	-2.74	2.74	2.97	-8.8	$104.N_2O_4$
UCCSD(T*)-INT-F12/A	1.25	1.37	1.80	7.3	$103.N_2O_3$
UCCSD(T*)-INT-F12/B	0.00	0.79	1.06	5.0	$103.N_2O_3$

Table D.3: Small Basis: **cc-pVQZ**

Method	mean error	MAD	RMS	Max Error	Molecule
	cc-pVDZ-F12				
UCCSD(T)+F12/A	16.66	16.66	17.86	33.9	$104.N_2O_4$
UCCSD(T)+F12/B	1.38	1.66	1.98	4.3	$59.C_2N_2$
UCCSD(T*)+F12/A	18.98	18.98	20.30	41.5	$104.N_2O_4$
UCCSD(T*)+F12/B	2.42	2.52	2.94	7.7	$104.N_2O_4$
UCCSD(T)-INT-F12/A	7.20	7.21	7.97	14.5	$103.N_2O_3$
UCCSD(T)-INT-F12/B	-2.97	2.97	3.12	-6.4	$104.N_2O_4$
UCCSD(T*)-INT-F12/A	9.52	9.52	10.43	22.1	$104.N_2O_4$
UCCSD(T*)-INT-F12/B	-1.93	1.93	2.07	-4.3	$5.CF_4$
	cc-pVTZ-F12				
UCCSD(T)+F12/A	9.51	9.51	9.99	16.7	$104.N_2O_4$
UCCSD(T)+F12/B	4.25	4.25	4.50	7.2	$31.CH_4N_2O$
UCCSD(T*)+F12/A	11.25	11.25	11.81	22.3	$104.N_2O_4$
UCCSD(T*)+F12/B	5.58	5.58	5.88	11.3	$104.N_2O_4$
UCCSD(T)-INT-F12/A	2.55	2.55	2.81	5.4	$39.C_2HF_3$
UCCSD(T)-INT-F12/B	-0.81	0.82	0.99	-3.8	$104.N_2O_4$
UCCSD(T*)-INT-F12/A	4.29	4.29	4.63	9.8	$103.N_2O_3$
UCCSD(T*)-INT-F12/B	0.52	0.63	0.80	3.0	$103.N_2O_3$
	cc-pVQZ-F12				
UCCSD(T)+F12/A	6.29	6.29	6.62	10.8	$31.CH_4N_2O$
UCCSD(T)+F12/B	4.58	4.59	4.84	7.8	$31.CH_4N_2O$
UCCSD(T*)+F12/A	7.84	7.84	8.23	15.5	$104.N_2O_4$
UCCSD(T*)+F12/B	5.99	5.99	6.29	11.8	$104.N_2O_4$
UCCSD(T)-INT-F12/A	0.53	0.69	0.85	2.0	$39.C_2HF_3$
UCCSD(T)-INT-F12/B	-0.55	0.60	0.82	-3.5	$104.N_2O_4$
UCCSD(T*)-INT-F12/A	2.08	2.08	2.32	6.0	$103.N_2O_3$
UCCSD(T*)-INT-F12/B	0.85	0.91	1.10	3.7	$103.N_2O_3$

Table D.4: Small Basis: **cc-pV5Z**

APPENDIX **E**

List of Scientific Publications

1. K. D. Vogiatzis, W. Klopper, A. Mavrandonakis, K. Fink,
   Magnetic properties of paddlewheels and trinuclear clusters with exposed metal sites,
   ChemPhysChem, **12**, 3307 (2011).

2. K. D. Vogiatzis, E. Barnes, W. Klopper,
   Interference-Corrected Explicitly-Correlated Second-Order Perturbation Theory,
   Chem. Phys. Lett., **503**, 157 (2011).

3. K. D. Vogiatzis, A. Mavrandonakis, W. Klopper, G. E. Froudakis,
   Ab initio study of the Interactions between CO_2 and N-Containing Organic Heterocycles,
   ChemPhysChem, **10**, 374 (2009).

APPENDIX F ▪

Acknowledgements

Steep was the path, and strewn with cobblestones...

The years that I spent in Karlsruhe and in the beautiful German nature will always be remembered by me. This journey gaved me priceless knowledge, both in a scientific and personal level, and filled me with unforgettable memories. But it would had never started and turned into a beautiful reality if some persons have not believed in me, have not supported me in the hard times and in the difficulties of this journey. In these next lines, I would like to thank them, one by one, for being there, next to me during this tough but beautiful period.

I would like to express my gratitude to my supervisor, Prof. Wim Klopper, for giving me a chance to work with him, for his excellent guidance, constant encouragement and criticism during the period of my doctorate. He stood to me like a second father to me all these years.

I am thankful to Priv.-Doz. Karin Fink for her supervision and collaboration that we had during my period at the Institute of Nanotechnology.

I would also like to thank Dr. Andreas Mavrantonakis, my first teacher in Crete and co-supervisor during the first months of my stay in Karlsruhe. It was not only his scientific guidance, but also his brilliant character that made a strong and valuable friendship to grow between us.

I don't know how should I show my gratitude to Dr. Robin Haunschild for the collaboration, the countless discussions and the knowledge he passed to me through them the last two years of my doctorate studies.

I am thankful also to Sandra Ahnen, Andrew Atkins, Dr. Robert Barthel, Dr. Angela Bihlmeier, Tilmann Bodenstein, Dr. Michael Harding, Konstantis Konidaris, Dr. Adam Kubas, Michael Kühn, Nils Middendorf, Angelos Polyzoidis and Dr. Andrey Yachmenev for the exciting, unique and monumental moments we shared together, in Karlsruhe and not only! Most especially to my colleagues, past or current members of our group, for creating a pleasant working environment at the university.

I would also like to thank my parents Ileana and Dionisis for providing me with the support needed in order to continually push myself to succeed, my godmother Magda for helping me confront with the difficulties that appeared all these years and my siblings Marilena and Manos for their love. Μου λείψατε πολύ όλα αυτά τα χρόνια.

Finally, I would like to thank my best friends Ioanna, Kostas and Orestis for their support all this period, in the good but also in the difficult times, even if they were so far away. You are very special to me.

The time I spent in Germany would not have been so exciting if I have not met Michael, a true friend that supported and inspired me for becoming a better man and scientist.

Last, but not least, a special word to my dear Olga. This endaviour would never have been finished without your love, your smile and constant support through the last year of my doctorate. Thank you for being there for me.

Special thanks: my Roubaix 🚲, Naturpark Schwarzwald (Nord/Mitte), the Reanimators and the Dawnrazors.

Bibliography

[1] Pople, J. A. *Angew. Chem. Int. Ed.* **1999**, *38*(13-14), 1894–1902.

[2] Bartlett, R. J.; Musial, M. *Rev. Mod. Phys.* **2007**, *79*(1), 291–352.

[3] Raghavachari, K.; Trucks, G. W.; Pople, J. A.; Head-Gordon, M. *Chem. Phys. Lett.* **1989**, *157*(6), 479–483.

[4] Watts, J. D.; Gauss, J.; Bartlett, R. J. *J. Chem. Phys.* **1993**, *98*(11), 8718–8733.

[5] Stanton, J. F. *Chem. Phys. Lett.* **1997**, *281*(1-3), 130–134.

[6] Hättig, C.; Klopper, W.; Köhn, A.; Tew, D. P. *Chem. Rev.* **2012**, *112*(1), 4–74.

[7] Klopper, W.; Bak, K. L.; Jørgensen, P.; Olsen, J.; Helgaker, T. *J. Phys. B: At. Mol. Opt. Phys.* **1999**, *32*(13), R103–R130.

[8] Tew, D. P.; Klopper, W.; Neiss, C.; Hättig, C. *Phys. Chem. Chem. Phys.* **2007**, *9*(16), 1921–1930.

[9] Knizia, G.; Adler, T. B.; Werner, H.-J. *J. Chem. Phys.* **2009**, *130*(5), 054104.

[10] Haunschild, R.; Klopper, W. *J. Chem. Phys.* **2012**, *136*(16), 164102.

[11] Vogiatzis, K. D.; Mavrandonakis, A.; Klopper, W.; Froudakis, G. E. *ChemPhysChem* **2009**, *10*(2), 374–383.

[12] Sillar, K.; Hofmann, A.; Sauer, J. *J. Am. Chem. Soc.* **2009**, *131*(11), 4143–4150.

[13] Nyden, M. R.; Petersson, G. A. *J. Chem. Phys.* **1981**, *75*(4), 1843–1862.

[14] Helgaker, T.; Jørgensen, P.; Olsen, J. *Molecular Electronic Structure Theory;* Wiley: Chichester, 2000.

[15] Schwartz, C. *Phys. Rev.* **1962**, *126*(3), 1015–1019.

[16] Schwartz, C. In *Methods in Computational Physics;* Alder, B., Fernbach, S., Rotenberg, M., Eds., Vol. 2; Academic Press, New York, 1963; pages 262–265.

[17] Hill, R. N. *J. Chem. Phys.* **1985**, *83*(3), 1173–1196.

[18] Kutzelnigg, W.; Morgan, J. D. *J. Chem. Phys.* **1992**, *96*(6), 4484–4508.

[19] Dunning, T. H. *J. Chem. Phys.* **1989**, *90*(2), 1007–1024.

[20] Kendall, R. A.; Dunning, T. H.; Harrison, R. H. *J. Chem. Phys.* **1992**, *96*(9), 6796–6806.

[21] Woon, D. E.; Dunning, T. H. *J. Chem. Phys.* **1993**, *98*(2), 1358–1371.

[22] Woon, D. E.; Dunning, T. H. *J. Chem. Phys.* **1995**, *103*(11), 4572–4586.

[23] Carroll, D. P.; Silverstone, H. J.; Metzger, R. M. *J. Chem. Phys.* **1979**, *71*(10), 4142–4161.

[24] Helgaker, T.; Klopper, W.; Tew, D. *Mol. Phys.* **2008**, *106*(16-18), 2107–2143.

[25] Kutzelnigg, W. *Phys. Chem. Chem. Phys.* **2008**, *10*(23), 3460–3468.

[26] Martin, J. M. L. *Chem. Phys. Lett.* **1996**, *259*(5-6), 669–678.

[27] Martin, J. M. L. *Theor. Chem. Acc.* **1997**, *97*(1-4), 227–231.

[28] Martin, J. M. L.; Taylor, P. R. *J. Chem. Phys.* **1997**, *106*(20), 8620–8623.

[29] Wilson, A. K.; Dunning, T. H. *J. Chem. Phys.* **1997**, *106*(21), 8718–8716.

[30] Helgaker, T.; Klopper, W.; Koch, H.; Noga, J. *J. Chem. Phys.* **1997**, *106*(23), 9639–9646.

[31] Halkier, A.; Helgaker, T.; Jørgensen, P.; Klopper, W.; Koch, H.; Olsen, J.; Wilson, A. K. *Chem. Phys. Lett.* **1998**, *286*(3-4), 243–252.

[32] Schwenke, D. W. *J. Chem. Phys.* **2005**, *122*(1), 014107.

[33] Bakowies, D. *J. Chem. Phys.* **2007**, *127*(8), 84105–84128.

[34] Feller, D. *J. Chem. Phys.* **1992**, *96*(8), 6104–6114.

[35] Peterson, K. A.; Woon, D. E.; Dunning, T. H. *J. Chem. Phys.* **1994**, *100*(10), 7410–7415.

[36] Feller, D.; Peterson, K. A.; Hill, J. G. *J. Chem. Phys.* **2011**, *135*(4), 44102–44119.

[37] Petersson, G. A.; Nyden, M. R. *J. Chem. Phys.* **1081**, *75*(7), 3423–3425.

[38] Petersson, G. A.; Bennett, A.; Tensfeldt, T. G.; Al-Laham, M. A.; Shirley, W. A.; Mantzaris, J. *J. Chem. Phys.* **1988**, *89*(4), 2193–2218.

[39] Szabo, A.; Ostlund, N. S. *Modern Quantum Chemistry;* Dover: New York, 1982.

[40] Petersson, G. A.; Licht, S. L. *J. Chem. Phys.* **1981**, *75*(9), 4556–4566.

[41] Vogiatzis, K. D.; Barnes, E.; Klopper, W. *Chem. Phys. Lett.* **2011**, *503*(1-3), 157–161.

[42] Pople, J. A. In *Energy, Structure and Reactivity*; Smith, D. W., Mcrae, W. B., Eds.; Wiley: New York, 1973; page 51.

[43] Petersson, G. A.; Al-Laham, M. A. *J. Chem. Phys.* **1991**, *94*(9), 6081–6090.

[44] Petersson, G. A.; Tensfeldt, T. G.; Montgomery, J. A. *J. Chem. Phys.* **1991**, *94*(9), 6091–6101.

[45] Montgomery, J. A.; Ochterski, J. W.; Petersson, G. A. *J. Chem. Phys.* **1994**, *101*(7), 5900–5909.

[46] Ochterski, J. W.; Petersson, G. A.; Montgomery, J. A. *J. Chem. Phys.* **1995**, *104*(7), 2598–5619.

[47] Montgomery, J. A.; Frisch, M. J.; Ochterski, J. W.; Petersson, G. A. *J. Chem. Phys.* **1999**, *110*(6), 2822–2827.

[48] Montgomery, J. A.; Frisch, M. J.; Ochterski, J. W.; Petersson, G. A. *J. Chem. Phys.* **2000**, *112*(15), 6532–6542.

[49] Wood, G. P. F.; Random, L.; Petersson, G. A.; Barnes, E. C.; Frisch, M. J.; John A. Montgomery, J. *J. Chem. Phys.* **2006**, *125*(9), 94106–94121.

[50] Pople, J. A.; Head-Gordon, M.; Raghavachari, K. *J. Chem. Phys.* **1987**, *87*(10), 5968–5974.

[51] Petersson, G. A.; Yee, A. K.; Bennett, A. *J. Chem. Phys.* **1985**, *83*(10), 5105–5128.

[52] Pople, J. A.; Head-Gordon, M.; Fox, D. J.; Raghavachari, K.; Curtiss, L. A. *J. Chem. Phys.* **1989**, *90*(10), 5622–5629.

[53] Curtiss, L. A.; Raghavachari, K.; Trucks, G. W.; Pople, J. A. *J. Chem. Phys.* **1991**, *94*(11), 7221–7230.

[54] Curtiss, L. A.; Raghavachari, K.; Redfern, P. C.; Rassolov, V.; Pople, J. A. *J. Chem. Phys.* **1998**, *109*(18), 7764–7776.

[55] Curtiss, L. A.; Redfern, P. C.; Raghavachari, K. *J. Chem. Phys.* **2007**, *126*(8), 084108.

[56] Martin, J. M. L.; de Oliveira, G. *J. Chem. Phys.* **1999**, *111*(5), 1843–1856.

[57] Boese, A. D.; Oren, M.; Atasoylu, O.; Martin, J. M. L.; Kállay, M.; Gauss, J. *J. Chem. Phys.* **2004**, *120*(9), 4129–4141.

[58] Karton, A.; Daon, S.; Martin, J. M. L. *Chem. Phys. Lett.* **2011**, *510*(4-6), 165–178.

[59] Karton, A.; Martin, J. M. L. *J. Chem. Phys.* **2012**, *136*(12), 124114.

[60] Tajti, A.; Szalay, P. G.; Császár, A. G.; Kállay, M.; Gauss, J.; Valeev, E. F.; Flowers, B. A.; Vázquez, J.; Stanton, J. F. *J. Chem. Phys.* **2004**, *121*(23), 11599–11613.

[61] Bomble, Y. J.; Vázquez, J.; Kállay, M.; Michauk, C.; Szalay, P. G.; Császár, A. G.; Gauss, J.; Stanton, J. F. *J. Chem. Phys.* **2006**, *125*(6), 064108.

[62] Harding, M. E.; Vázquez, J.; Ruscic, B.; Wilson, A. K.; Gauss, J.; Stanton, J. F. *J. Chem. Phys.* **2008**, *128*(11), 114111.

[63] DeYonker, N. J.; Cundari, T. R.; Wilson, A. K. *J. Chem. Phys.* **2006**, *124*(11), 114104.

[64] DeYonker, N. J.; Peterson, K. A.; Steyl, G.; Wilson, A. K.; Cundari, T. R. *J. Phys. Chem. A* **2007**, *111*(44), 11269–11277.

[65] DeYonker, N. J.; Williams, T. G.; Imel, A. E.; Cundari, T. R.; Wilson, A. K. *J. Chem. Phys.* **2009**, *131*(2), 024106.

[66] Nedd, S. A.; DeYonker, N. J.; Wilson, A. K.; Piecuch, P.; Gordon, M. S. *J. Chem. Phys.* **2012**, *136*(14), 144109.

[67] Barnes, E. C.; Petersson, G. A.; Montgomery, J. A.; Frisch, M. J.; Martin, J. M. L. *J. Chem. Theory Comput.* **2009**, *5*(10), 2687–2693.

[68] Aguilera-Iparraguirre, J.; Curran, H. J.; Klopper, W.; Simmie, J. M.

[69] Peterson, K. A.; Feller, D.; Dixon, D. A. *Theor. Chem. Acc.* **2012**, *131*(1), 1079–1098.

[70] Møller, C.; Plesset, M. S. *Phys. Rev.* **1934**, *46*(7), 618–622.

[71] Tew, D. P.; Klopper, W.; Helgaker, T. *J. Comput. Chem.* **2007**, *28*(8), 1307–1320.

[72]   J.Roothaan, C. C. *Rev. Mod. Phys.* **1951**, *23*(2), 69–89.

[73]   Kato, T. *Commun. Pure Appl. Math.* **1957**, *10*(2), 151–177.

[74]   Pack, R. T.; Byers-Brown, W. *J. Chem. Phys.* **1966**, *45*(2), 556–559.

[75]   Tew, D. P. *J. Chem. Phys.* **2008**, *129*(1), 014104.

[76]   Slater, J. C. *Phys. Rev.* **1928**, *31*(3), 333–343.

[77]   Hylleraas, E. A. *Z. Physik* **1929**, *54*(5-6), 347–366.

[78]   Kong, L.; Bischoff, F. A.; Valeev, E. F. *Chem. Rev.* **2012**, *112*(1), 75–107.

[79]   Szalewicz, K.; Jeziorski, B.; Monkhorst, H. J.; Zabolitzky, J. G. *Chem. Phys. Lett.* **1982**, *91*(3), 169–172.

[80]   Szalewicz, K.; Zabolitzky, J. G.; Jeziorski, B.; Monkhorst, H. J. *J. Chem. Phys.* **1984**, *81*(6), 2723–2731.

[81]   Tew, D. P.; Klopper, W.; Manby, F. R. *J. Chem. Phys.* **2007**, *127*(17), 174105.

[82]   Kutzelnigg, W. *Theor. Chim. Acta* **1985**, *68*(6), 445–469.

[83]   Klopper, W.; Kutzelnigg, W. *Chem. Phys. Lett.* **1987**, *134*(1), 17–22.

[84]   Ten-no, S. *Chem. Phys. Lett.* **2000**, *330*(1-2), 169–174.

[85]   Klopper, W.; Samson, C. C. M. *J. Chem. Phys.* **2002**, *116*(15), 6397.

[86]   Valeev, E. F. *Chem. Phys. Lett.* **2004**, *395*(4-6), 190–195.

[87]   Tew, D. P.; Hättig, C.; Bachorz, R. A.; Klopper, W. In *Recent Progress in Coupled-Cluster Methods - Theory and Applications*; Čársky, P., Paldus, J., Pittner, J., Eds.; Springer: Dordrecht, Heidelberg, London, New York, 2010; pages 535–572.

[88]   Fliegl, H.; Klopper, W.; Hättig, C. *J. Chem. Phys.* **2005**, *122*(8), 084107.

[89]   Fliegl, H.; Hättig, C.; Klopper, W. *Int. J. Quant. Chem.* **2006**, *106*(11), 2306–2317.

[90]   Ahlrichs, R.; Bär, M.; Häser, M.; Horn, H.; Kölmel, C. *Chem. Phys. Lett.* **1989**, *162*(3), 165–169.

[91]   Löwdin, P.-O. *Phys. Rev.* **1955**, *97*(6), 1474–1489.

[92]   Löwdin, P.-O.; Shull, H. *Phys. Rev.* **1956**, *101*(6), 1730.

[93]   Edmiston, C.; Krauss, M. *J. Chem. Phys.* **1965**, *42*(3), 1119.

[94]   Tew, D. P.; Helmich, B.; Hättig, C. *J. Chem. Phys.* **2011**, *135*(7), 074107.

[95]   Neese, F.; Wennmohs, F.; Hansen, A. *J. Chem. Phys.* **2009**, *130*(11), 114108.

[96]   Neese, F.; Hansen, A.; Liakos, D. G. *J. Chem. Phys.* **2009**, *131*(6), 064103.

[97]   Liakos, D. G.; Hansen, A.; Neese, F. *J. Chem. Theory Comput.* **2011**, *7*(1), 76–87.

[98]   Yang, J.; Kurashige, Y.; Manby, F. R.; Chan, G. K. L. *J. Chem. Phys.* **2011**, *134*(4), 044123.

[99] Kurashige, Y.; Yang, J.; Chan, G. K. L.; Manby, F. R. *J. Chem. Phys.* **2012**, *136*(12), 124106.

[100] Yang, J.; Chan, G. K.-L.; Manby, F. R.; Schütz, M.; Werner, H.-J. *J. Chem. Phys.* **2012**, *136*(14), 144105.

[101] Anderson, E.; Bai, Z.; Bischof, C.; Blackford, S.; Demmel, J.; Dongarra, J.; Du Croz, J.; Greenbaum, A.; Hammarling, S.; McKenney, A.; Sorensen, D. *LAPACK Users' Guide;* Society for Industrial and Applied Mathematics: Philadelphia, PA, third ed., 1999.

[102] Bachorz, R. A.; Bischoff, F. A.; Glöss, A.; Hättig, C.; Höfener, S.; Klopper, W.; Tew, D. P. *J. Comput. Chem.* **2011**, *32*(11), 2492–2513.

[103] Peterson, K. A.; Adler, T. B.; Werner, H.-J. *J. Chem. Phys.* **2008**, *128*(8), 84102–84114.

[104] Pulay, P. *Chem. Phys. Lett.* **1983**, *100*(2), 151–154.

[105] Boys, S. F. *Rev. Mod. Phys.* **1960**, *32*(2), 296–299.

[106] Foster, J. M.; Boys, S. F. *Rev. Mod. Phys.* **1960**, *32*(2), 300–302.

[107] Pipek, J.; Mezey, P. G. *J. Chem. Phys.* **1989**, *90*(9), 4916–4926.

[108] Azar, R. J.; Head-Gordon, M. *J. Chem. Phys.* **2012**, *136*(2), 024103.

[109] Köhn, A.; Tew, D. P. *J. Chem. Phys.* **2010**, *132*(2), 024101.

[110] J.Roothaan, C. C.; Sachs, L. M.; Weiss, A. W. *Rev. Mod. Phys.* **1960**, *32*(2), 186–194.

[111] Barnes, E. C.; Petersson, G. A.; Feller, D.; Peterson, K. A. *J. Chem. Phys.* **2008**, *129*(19), 2008.

[112] Klopper, W. *Mol. Phys.* **2001**, *99*(6), 481–507.

[113] Flores, J. R. *Int. J. Quant. Chem.* **2008**, *108*(12), 2172–2177.

[114] Klopper, W.; Ruscic, B.; Tew, D. P.; Bischoff, F. A.; Wolfsegger, S. *Chem. Phys.* **2009**, *356*(1-3), 14–24.

[115] Grimme, S. *J. Chem. Phys.* **2003**, *118*(20), 9095–9102.

[116] Neese, F.; Valeev, E. F. *J. Chem. Theory Comput.* **2011**, *7*(1), 33–43.

[117] Samson, C. C. M.; Klopper, W. *Mol. Phys.* **2004**, *102*(23-24), 2499–2510.

[118] Jensen, F. *Introduction to Computational Chemistry;* Wiley: Chichester, 2007.

[119] Yousaf, K. E.; Peterson, K. A. *J. Chem. Phys.* **2008**, *129*(18), 184108–184115.

[120] Hättig, C. *Phys. Chem. Chem. Phys.* **2005**, *7*(1), 59–66.

[121] Weigend, F. *J. Comput. Chem.* **2008**, *29*(2), 167–175.

[122] Middendorf, N.; Höfener, S.; Klopper, W.; Helgaker, T. *Chem. Phys.* **2012**, *401*(1), 146–151.

[123] Papajak, E.; Truhlar, D. G. *J. Chem. Phys.* **2012**, *137*(6), 064110.

[124] Jurečka, P.; Hobza, P. *Chem. Phys. Lett.* **2002**, *365*(1-2), 89–94.

[125] Hobza, P.; Šponer, J. *J. Am. Chem. Soc.* **2002**, *124*(39), 11802–11808.

[126] Jurečka, P.; Šponer, J.; Černý, J.; Hobza, P. *Phys. Chem. Chem. Phys.* **2006**, *8*(17), 1985–1993.

[127] Lynch, B. J.; Truhlar, D. G. *J. Phys. Chem. A* **2003**, *107*(42), 8996–8999.

[128] Curtiss, L. A.; Raghavachari, K.; Redfern, P. C.; Pople, J. A. *J. Chem. Phys.* **1997**, *106*(3), 1063–1079.

[129] Curtiss, L. A.; Redfern, P. C.; Raghavachari, K.; Pople, J. A. *J. Chem. Phys.* **1998**, *109*(1), 42–54.

[130] Unpublished results obtained from active thermochemical tables (atct) ver. 1.36 using the core (argonne) thermochemical network ver. 1.070. Ruscic, B.; Core (Argonne) Thermochemical Network, **2008**.

[131] Ruscic, B.; Pinzon, R. E.; Morton, M. L.; von Laszevski, G.; Bittner, S. J.; Nijsure, S. G.; Amin, K. A.; Minkoff, M.; Wagner, A. F. *J. Phys. Chem. A* **2004**, *108*(45), 9979–9997.

[132] Klopper, W.; Bachorz, R. A.; Hättig, C.; Tew, D. P. *Theor. Chem. Acc.* **2010**, *126*(5-6), 289–304.

[133] Bachorz, R. *Implementation and Application of the Explicitly-Correlated Coupled-Cluster Method in Turbomole* PhD thesis, Universität Karlsruhe, **2009**.

[134] Weigend, F.; Ahlrichs, R. *Phys. Chem. Chem. Phys.* **2005**, *7*(18), 3297–3305.

[135] Hill, J. G.; Mazumder, S.; Peterson, K. A. *J. Chem. Phys.* **2010**, *132*(5), 054108.

[136] Ten-no, S. *J. Chem. Phys.* **2004**, *121*(1), 117–129.

[137] Bokhan, D.; Ten-no, S.; Noga, J. *Phys. Chem. Chem. Phys.* **2008**, *10*(23), 3320–3326.

[138] Tew, D. P.; Klopper, W. *Mol. Phys.* **2010**, *108*(3-4), 315–325.

[139] Marchetti, O.; Werner, H.-J. *J. Phys. Chem. A* **2009**, *113*(43), 11580–11585.

[140] Marshall, M. S.; Sherrill, C. D. *J. Chem. Theory Comput.* **2011**, *7*(12), 3978–3982.

[141] Haunschild, R.; Klopper, W. *Theor. Chem. Acc.* **2012**, *131*(2), 1112–1118.

[142] Janssen, C. L.; Nielsen, I. M. *Chem. Phys. Lett.* **1998**, *290*(4-6), 423–430.

[143] Lynch, B. J.; Truhlar, D. G. *J. Phys. Chem. A* **2003**, *107*(19), 3898–3906.

[144] Lynch, B. J.; Zhao, Y.; Truhlar, D. G. *J. Phys. Chem. A* **2005**, *109*(8), 1643–1649.

[145] Karton, A.; Rabinovich, E.; Martin, J. M. L.; Ruscic, B. *J. Chem. Phys.* **2006**, *125*(14), 144108.

[146] Dunning, T. H.; Peterson, K. A.; Wilson, A. K. *J. Chem. Phys.* **2001**, *114*(21), 9244–9253.

[147] Curtiss, L. A.; Raghavachari, K.; Redfern, P. C.; Pople, J. A. *J. Chem. Phys.* **2000**, *112*(17), 7374–7373.

[148] Haunschild, R.; Janesko, B. G.; Scuseria, G. E. *J. Chem. Phys.* **2009**, *131*(15), 154112.

[149] Müller-Dethlefs, K.; Hobza, P. *Chem. Rev.* **2000**, *100*(1), 143–168.

[150] Hobza, P. *Acc. Chem. Res.* **2012**, *45*(4), 663–672.

[151] Li, J.-R.; Sculley, J.; Zhou, H.-C. *Chem. Rev.* **2012**, *112*(2), 869–932.

[152] Harlick, P. J. E.; Tezel, F. H. *Microp. Mesop. Mater.* **2004**, *76*(1-3), 71–79.

[153] Himeno, S.; Komatsu, T.; Fujita, S. *J. Chem. Eng. Data* **2005**, *50*(2), 369–376.

[154] Sumida, K.; Rogow, D. L.; Mason, J. A.; McDonald, T. M.; Bloch, E. D.; Herm, Z. R.; Bae, T.-H.; Long, J. R. *Chem. Rev.* **2012**, *112*(2), 724–781.

[155] Suh, M. P.; Park, H. J.; Prasad, T. K.; Lim, D.-W. *Chem. Rev.* **2012**, *112*(2), 782–835.

[156] Schlapbach, L.; Züttel, A. *Nature* **2001**, *414*(6861), 353–358.

[157] Cohen, S. M. *Chem. Rev.* **2012**, *112*(2), 970–1000.

[158] Lehn, J.-M. *Science* **1993**, *260*(5115), 1762–1763.

[159] Grimme, S. *J. Comput. Chem.* **2004**, *25*(12), 1463–1473.

[160] Grimme, S. *J. Comput. Chem.* **2006**, *27*(15), 1787–1799.

[161] Grimme, S.; Antony, J.; Ehrlich, S.; Krieg, H. *J. Chem. Phys.* **2010**, *132*(15), 154104.

[162] Torres, E.; DiLabio, G. A. *J. Phys. Chem. Lett.* **2012**, *3*(12), 1738–1744.

[163] Ehrlich, S.; Moellmann, J.; Grimme, S. *Acc. Chem. Res.* **2012**, *???*(???), ???

[164] Stone, A. J. *Science* **2008**, *321*(5890), 787–789.

[165] Buckingham, A. D.; Fowler, P. W. *Can. J. Chem.* **1985**, *63*(7), 2018–2025.

[166] Stone, A. J. *The Theory of Intermolecular Forces;* Oxford University Press: Oxford, 1996.

[167] Riley, K. E.; Pitoňák, M.; Jurečka, P.; Hobza, P. *Chem. Rev.* **2010**, *110*(9), 5023–5063.

[168] Řezáč, J.; Riley, K. E.; Hobza, P. *J. Chem. Theory Comput.* **2011**, *7*(8), 2427–2438.

[169] Řezáč, J.; Riley, K. E.; Hobza, P. *J. Comput. Chem.* **2012**, *33*(6), 691–694.

[170] Boys, S.; Bernardi, F. *Mol. Phys.* **1970**, *19*(4), 553–566.

[171] Vogiatzis, K. D. Theoretical study of interactions between carbon dioxide and nanomaterials Master's thesis, University of Crete, **2008**.

[172] Jung, Y.; Lochan, R. C.; Dutoi, A. D.; Head-Gordon, M. *J. Chem. Phys.* **2004**, *121*(20), 9793.

[173] Park, K. S.; Ni, Z.; Côté, A. P.; Choi, J. Y.; Huang, R.; Uribe-Romo, F. J.; Chae, H. K.; O'Keeffe, M.; Yaghi, O. M. *Proc. Natl. Acad. Sci.* **2006**, *103*(27), 10186–10191.

[174] Doran, J. L.; Hon, B.; Leopold, K. R. *J. Mol. Struct.* **2012**, *1019*(1), 191–195.

[175] Phillips, J. A.; Canagaratna, M.; Goodfriend, H.; Grushow, A.; Almlöf, J.; Leopold, K. R. *J. Am. Chem. Soc.* **1995**, *117*(50), 12549–12556.

[176] Pitoňák, M.; Neogrády, P.; Černý, J.; Grimme, S.; Hobza, P. *ChemPhysChem* **2009**, *10*(1), 282–289.

[177] Deshmukh, M. M.; Sakaki, S. *Theor. Chem. Acc.* **2011**, *130*(2-3), 475–482.

[178] deLange, K. M.; Lane, J. R. *J. Chem. Phys.* **2011**, *134*(3), 034301.

[179] deLange, K. M.; Lane, J. R. *J. Chem. Phys.* **2011**, *135*(6), 064304.

[180] Mackie, I. D.; DiLabio, G. A. *Phys. Chem. Chem. Phys.* **2011**, *13*(7), 2780–2787.

[181] An, J.; Geib, S.; Rosi, N. L. *J. Am. Chem. Soc.* **2010**, *132*(1), 38–39.

[182] Babarao, R.; Dai, S.; en Jiang, D. *Langmuir* **2011**, *27*(7), 3451–3460.

[183] Pachfule, P.; Das, R.; Poddar, P.; Banerjee, R. *Cryst. Growth Des.* **2010**, *10*(6), 2475–2478.

[184] Panda, T.; Pachfule, P.; Chen, Y.; Jiang, J.; Banerjee, R. *Chem. Comm.* **2011**, *47*(7), 2011–2013.

[185] Si, X.; Jiao, C.; Li, F.; Zhang, J.; Wang, S.; Liu, S.; Li, Z.; Sun, L.; Xu, F.; Gabelica, Z.; Schick, C. *Energy Environ. Sci.* **2011**, *4*(11), 4522–4527.

[186] Qin, J.-S.; Du, D.-Y.; Li, W.-L.; Zhang, J.-P.; Li, S.-L.; Su, Z.-M.; Wang, X.-L.; Xu, Q.; Shao, K.-Z.; Lan, Y.-Q. *Chem. Sci.* **2012**, *3*(6), 2114–2118.

[187] Vitillo, J. G.; Savonneta, D. M.; Ricchiardi, G.; Bordiga, S. *ChemSusChem* **2011**, *4*(9), 1281–1290.

[188] Du, N.; Park, H. B.; Robertson, G. P.; Dal-Cin, M. M.; Visser, T.; Scoles, L.; Guiver, M. D. *Nature Materials* **2011**, *10*(5), 372–375.

[189] Park, T.-H.; Cychosz, K. A.; Wong-Foy, A. G.; Daillyb, A.; Matzger, A. J. *Chem. Comm.* **2011**, *47*(5), 1452–1454.

[190] Rablen, P. R.; Lockman, J. W.; Jorgensen, W. L. *J. Phys. Chem. A* **1998**, *102*(21), 3782–3797.

[191] Mascal, M.; Armstrong, A.; Bartberger, M. D. *J. Am. Chem. Soc.* **2001**, *124*(22), 6274–6276.

[192] Zhao, Y.; Truhlar, D. G. *J. Chem. Theory Comput.* **2006**, *2*(4), 1009–1018.

[193] Patkowski, K. *J. Chem. Phys.* **2012**, *137*(3), 034103.

[194] Tew, D. P.; Klopper, W. *J. Chem. Phys.* **2006**, *135*(9), 094302.

[195] Werner, H.-J.; Adler, T. B.; Manby, F. R. *J. Chem. Phys.* **2007**, *126*(16), 164102.

[196] Burda, J. V.; Zahradnik, R.; Hobza, P.; Urban, M. *Mol. Phys.* **1996**, *89*(2), 425–432.

[197] Aziz, R.; Meath, W. J.; Allnatt, A. *Chem. Phys.* **1983**, *78*(2), 295–309.

[198] Saebø, S.; Pulay, P. *Annu. Rev. Phys. Chem.* **1993**, *44*, 213–236.

[199] Paul, A.; Kubicki, M.; Kubas, A.; Jelsch, C.; Fink, K.; Lecomte, C. *J. Phys. Chem. A* **2011**, *115*(45), 12941–12952.

[200] Riley, K. E.; Platts, J. A.; Řezáč, J.; Hobza, P.; Hill, J. G. *J. Phys. Chem. A* **2012**, *116*(16), 4159–4169.

[201] Jurečka, P.; Hobza, P. *J. Am. Chem. Soc.* **2003**, *125*(50), 15608–15613.

[202] Šponer, J.; Jurečka, P.; Hobza, P. *J. Am. Chem. Soc.* **2004**, *126*(32), 10142–10151.

[203] Sato, T.; Tsuneda, T.; Hirao, K. *J. Chem. Phys.* **2007**, *126*(23), 234114.

[204] Schwabe, T.; Grimme, S. *Phys. Chem. Chem. Phys.* **2007**, *9*(26), 3397–3406.

[205] Chai, J.-D.; Head-Gordon, M. *Phys. Chem. Chem. Phys.* **2008**, *10*(44), 6615–6620.

[206] Lee, K.; Murray, É. D.; Kong, L.; Lundqvist, B. I.; Langreth, D. C. *Phys. Rev. B* **2010**, *82*(8), 081101.

[207] Arabi, A. A.; Becke, A. D. *J. Chem. Phys.* **2012**, *137*(1), 014104.

[208] Podeszwa, R.; Patkowski, K.; Szalewicz, K. *Phys. Chem. Chem. Phys.* **2010**, *12*(23), 5974–5979.

[209] Takatani, T.; Hohenstein, E. G.; Malagoli, M.; Marshall, M. S.; Sherrill, C. D. *J. Chem. Phys.* **2010**, *132*(14), 144104.

[210] Grimme, S. *J. Chem. Phys.* **2006**, *124*(3), 034108.

[211] Distasio, R. A.; Head-Gordon, M. *Mol. Phys.* **2007**, *105*(8), 1073–1083.

[212] Marshall, M. S.; Burns, L. A.; Sherrill, C. D. *J. Chem. Phys.* **2011**, *135*(19), 194102.

[213] Marchetti, O.; Werner, H.-J. *Phys. Chem. Chem. Phys.* **2008**, *10*(23), 3400–3409.

[214] Wheeler, S. E. *Acc. Chem. Res.* **2012**.

[215] Karthikeyan, S.; Nagase, S. *J. Phys. Chem. A* **2012**, *116*(7), 1694–1700.

[216] Pfaffen, C.; Infanger, D.; Ottiger, P.; Frey, H.-M.; Leutwyler, S. *Phys. Chem. Chem. Phys.* **2011**, *13*(31), 14110–14118.

[217] Rodham, D. A.; Suzuki, S.; Suenram, R. D.; Lovas, F. J.; Dasgupta, S.; Goddard, W. A.; Blake, G. A. *Nature* **1993**, *362*(6422), 735–737.

[218] Tsuzuki, S.; Honda, K.; Uchimaru, T.; Mikami, M.; Tanabe, K. *J. Am. Chem. Soc.* **2000**, *122*(46), 11450–11458.

[219] Vaupel, S.; Brutschy, B.; Tarakeshwar, P.; Kim, K. S. *J. Am. Chem. Soc.* **2006**, *128*(16), 5416–5426.

[220] Columberg, G.; Bauder, A. *J. Chem. Phys.* **1997**, *106*(2), 504–510.

[221] Matsumoto, Y.; Honma, K. *J. Chem. Phys.* **2007**, *127*(18), 184310.

[222] Dauster, I.; Rice, C. A.; Zielke, P.; Suhm, M. A. *Phys. Chem. Chem. Phys.* **2008**, *10*(19), 2827–2835.

[223] Ottiger, P.; Pfaffen, C.; Leist, R.; Leutwyler, S.; Bachorz, R. A.; Klopper, W. *J. Phys. Chem. B* **2009**, *113*(9), 2937–2943.

[224] Pfaffen, C.; Frey, H.-M.; Ottiger, P.; Leutwyler, S.; Bachorz, R. A.; Klopper, W. *Phys. Chem. Chem. Phys.* **2010**, *12*(29), 8208–8218.

[225] Vogiatzis, K. D.; Ahnen, S.; Pfaffen, C.; Leutwyler, S.; Klopper, W. *In Preparation* **201X**.

[226] Ward, M. D. *Science* **2003**, *300*(5622), 1104–1105.

[227] Collins, D. J.; Zhou, H.-C. *J. Mater. Chem.* **2007**, *17*(30), 3154–3160.

[228] Murray, L. J.; Dincă, M.; Long, J. R. *Chem. Soc. Rev.* **2009**, *38*(5), 1294–1314.

[229] Férey, G. *Chem. Soc. Rev.* **2008**, *37*(1), 191–214.

[230] Kitagawa, S.; Kitaura, R.; ichiro Noro, S. *Angew. Chem. Int. Ed.* **2004**, *43*, 2334–2375.

[231] Tranchemontagne, D. J.; Mendoza-Cortés, J. L.; O'Keeffe, M.; Yaghi, O. M. *Chem. Soc. Rev.* **2009**, *38*(5), 1257–1283.

[232] Lee, J.; Farha, O. K.; Roberts, J.; Scheidt, K. A.; Nguyen, S. T.; Hupp, J. T. *Chem. Soc. Rev.* **2009**, *38*(5), 1450–1459.

[233] Li, J.-R.; Kuppler, R. J.; Zhou, H.-C. *Chem. Soc. Rev.* **2009**, *38*(5), 1477–1504.

[234] Maspoch, D.; Ruiz-Molina, D.; Wurst, K.; Domingo, N.; Cavallini, M.; Biscarini, F.; Tejada, J.; Rovira, C.; Veciana, J. *Nat. Mater.* **2003**, *2*(3), 190–195.

[235] Rosi, N. L.; Eckert, J.; Eddaoudi, M.; Vodak, D. T.; Kim, J.; O'Keeffe, M.; Yaghi, O. M. *Science* **2003**, *300*(5622), 1127–1129.

[236] Rowsell, J. L. C.; Eckert, J.; Yaghi, O. M. *J. Am. Chem. Soc.* **2005**, *127*(42), 14904–14910.

[237] Klontzas, E.; Mavrandonakis, A.; Froudakis, G.; Carissan, Y.; Klopper, W. *J. Phys. Chem. C* **2007**, *111*(36), 13635–13640.

[238] Kubas, G. J. *Chem. Rev.* **2007**, *107*(10), 4152–4205.

[239] Dincă, M.; Long, J. R. *Angew. Chem. Int. Ed.* **2008**, *47*(36), 6766–6779.

[240] Chui, S. S.-Y.; Lo, S. M.-F.; Charmant, J. P. H.; Orpen, A. G.; Williams, I. D. *Science* **1999**, *283*(5405), 1148–1150.

[241] Chen, B.; Ockwig, N. W.; Millward, A. R.; Contreras, D. S.; Yaghi, O. M. *Angew. Chem. Int. Ed.* **2005**, *44*(30), 4745–4749.

[242] Lin, X.; Jia, J.; Zhao, X.; Thomas, K. M.; Blake, A. J.; Walker, G. S.; Champness, N. R.; Hubberstey, P.; Schröder, M. *Angew. Chem. Int. Ed.* **2006**, *45*(44), 7358–7364.

[243] Sun, D.; Ma, S.; Ke, Y.; Collins, D. J.; Zhou, H.-C. *J. am. Chem. Soc.* **2006**, *128*(12), 3896–3897.

[244] Ma, S.; Sun, D.; Ambrogio, M.; Fillinger, J. A.; Parkin, S.; Zhou, H.-C. *J. Am. Chem. Soc.* **2007**, *129*(7), 1858–1859.

[245] Nouar, F.; Eubank, J. F.; Bousquet, T.; Wojtas, L.; Zaworotko, M. J.; Eddaoudi, M. *J. Am. Chem. Soc.* **2008**, *130*(6), 1833–1835.

[246] Lin, X.; Telepeni, I.; Blake, A. J.; Dailly, A.; Brown, C. M.; Simmons, J. M.; Zoppi, M.; Walker, G. S.; Thomas, K. M.; Mays, T. J.; Hubbersty, P.; Champness, N. R.; Schröder, M. *J. Am. Chem. Soc.* **2009**, *131*(6), 2159–2171.

[247] Farha, O. K.; Yazaydın, A. O.; Eryazici, I.; Malliakas, C. D.; Hauser, B. G.; Kanatzidis, M. G.; Nguyen, S. T.; Snurr, R. Q.; Hupp, J. T. *Nature Chem.* **2010**, *2*(11), 944–948.

[248] Loiseau, T.; Lecroq, L.; Volkringer, C.; Marrot, J.; Férey, G.; Haouas, M.; Taulelle, F.; Bourrelly, S.; Llewellyn, P. L.; Latroche, M. *J. Am. Chem. Soc.* **2006**, *128*(31), 10223–10230.

[249] Férey, G.; Mellot-Draznieks.; Serre, C.; Millange, F.; Dutour, J.; Surblé, S.; Margiolaki, I. *Science* **2005**, *309*(5743), 2040–2042.

[250] Latroche, M.; Surblé, S.; Serre, C.; Mellot-Draznieks, C.; Llewellyn, P. L.; Lee, J.-H.; Chang, J.-S.; Jhung, S. H.; Férey, G. *Angew. Chem. Int. Ed.* **2006**, *45*(48), 8227–8231.

[251] Surblé, S.; Millange, F.; Serre, C.; Düren, T.; Latroche, M.; Bourrelly, S.; Llewellyn, P. L.; Férey, G. *J. Am. Chem. Soc.* **2006**, *128*(46), 14889–14896.

[252] Yoon, J. H.; Choi, S. B.; Oh, Y. J.; Seo, M. J.; Jhon, Y. H.; Lee, T.-B.; Kim, D.; Choi, S. H.; Kim, J. *Catal. Today* **2007**, *120*(3-4), 324–329.

[253] Ma, S.; Wang, X.-S.; Manis, E. S.; Collier, C. D.; Zhou, H.-C. *Inorg. Chem.* **2007**, *46*(9), 3432–3434.

[254] Liu, Y.; Eubank, J.-F.; Cairns, A.-J.; Eckert, J.; Kravtsov, V.-C.; Luebke, R.; Eddaoudi, M. *Angew. Chem. Int. Ed.* **2007**, *46*(18), 3278–3283.

[255] Vogiatzis, K. D.; Klopper, W.; Mavrandonakis, A.; Fink, K. *ChemPhysChem* **2011**, *12*(17), 3307–3319.

[256] Bak, J. H.; Le, V.-D.; Kang, J.; Wei, S.-H.; Kim, Y.-H. *J. Phys. Chem. C* **2012**, *116*(13), 7386–392.

[257] Cannon, R. D.; White, R. In *Progress in Inorganic Chemistry*; Lippard, S. J., Ed., Vol. 36; Wiley, 1988; pages 195–298.

[258] Perdew, J. P.; Burke, K.; Ernzerhof, M. *Phys. Rev. Lett.* **1996**, *77*(18), 3865–3868.

[259] Weigend, F.; Häser, M.; Patzelt, H.; Ahlrichs, R. *Chem. Phys. Lett.* **1998**, *294*(1-3), 143–152.

[260] Becke, A. D. *Phys. Rev. A* **1988**, *38*(6), 3098–3100.

[261] Tao, J.; Perdew, J. P.; Staroverov, V. N.; Scuseria, G. E. *Phys. Rev. Lett.* **2003**, *91*(14), 146401–146404.

[262] Lee, C. T.; Yang, W. T.; Parr, R. G. 785-789 **1988**, *37*(2).

[263] Becke, A. D. *J. Chem. Phys* **1993**, *98*(7), 5648–5652.

[264] Nyden, M. R.; Petersson, G. A. *Int. J. Quant. Chem.* **1980**, *17*(5), 975–982.

ibidem-Verlag

Melchiorstr. 15

D-70439 Stuttgart

info@ibidem-verlag.de

www.ibidem-verlag.de
www.ibidem.eu
www.edition-noema.de
www.autorenbetreuung.de

www.ingramcontent.com/pod-product-compliance
Lightning Source LLC
Chambersburg PA
CBHW061926190326
41458CB00009B/2669